STUDIES

THE DIFFUSION OF ADVANCED TELECOMMUNICATIONS IN DEVELOPING COUNTRIES

BY
CHRISTIANO ANTONELLI

DEVELOPMENT CENTRE
OF THE ORGANISATION FOR ECONOMIC CO-OPERATION AND DEVELOPMENT

Pursuant to Article 1 of the Convention signed in Paris on 14th December 1960, and which came into force on 30th September 1961, the Organisation for Economic Co-operation and Development (OECD) shall promote policies designed:

— to achieve the highest sustainable economic growth and employment and a rising standard of living in Member countries, while maintaining financial stability, and thus to contribute to the development of the world economy;
— to contribute to sound economic expansion in Member as well as non-member countries in the process of economic development; and
— to contribute to the expansion of world trade on a multilateral, non-discriminatory basis in accordance with international obligations.

The original Member countries of the OECD are Austria, Belgium, Canada, Denmark, France, Germany, Greece, Iceland, Ireland, Italy, Luxembourg, the Netherlands, Norway, Portugal, Spain, Sweden, Switzerland, Turkey, the United Kingdom and the United States. The following countries became Members subsequently through accession at the dates indicated hereafter: Japan (28th April 1964), Finland (28th January 1969), Australia (7th June 1971) and New Zealand (29th May 1973). The Commission of the European Communities takes part in the work of the OECD (Article 13 of the OECD Convention). Yugoslavia takes part in some of the work of the OECD (agreement of 28th October 1961).

The Development Centre of the Organisation for Economic Co-operation and Development was established by decision of the OECD Council on 23rd October 1962.

The purpose of the Centre is to bring together the knowledge and experience available in Member countries of both economic development and the formulation and execution of general economic policies; to adapt such knowledge and experience to the actual needs of countries or regions in the process of development and to put the results at the disposal of the countries by appropriate means.

The Centre has a special and autonomous position within the OECD which enables it to enjoy scientific independence in the execution of its task. Nevertheless, the Centre can draw upon the experience and knowledge available in the OECD in the development field.

*
* *

THE OPINIONS EXPRESSED AND ARGUMENTS EMPLOYED IN THIS PUBLICATION ARE THE SOLE RESPONSIBILITY OF THE AUTHOR AND DO NOT NECESSARILY REFLECT THOSE OF THE OECD OR OF THE GOVERNMENTS OF ITS MEMBER COUNTRIES

Publié en français sous le titre :

LA DIFFUSION DES TÉLÉCOMMUNICATIONS DE POINTE
DANS LES PAYS EN DÉVELOPPEMENT

FOREWORD

This study was undertaken as part of the Development Centre's research project on New Technologies and Changing Comparative Advantages in Industry, under the direction of Charles Oman. It benefited from a grant from the Government of Italy, for which the Centre and the author express their gratitude.

ALSO AVAILABLE

Cut along dotted line

ORDER FORM

Please enter my order for:

Qty.	OECD Code	Title	Price
....	....	....	....
....	....	....	....
....	....	....	....
....	....	....	....
		Total :	....

- Payment is enclosed ☐
- Charge my VISA card ☐   Number of card ....
  (Note: You will be charged the French franc price.)
  Expiration of card .... *Signature* ....
- *Send invoice. A purchase order is attached* ☐

Send publications to *(please print)*:
  Name ....
  Address ....
  ....
  ....

Send this Order Form to OECD Publications Service, 2, rue André-Pascal, 75775 PARIS CEDEX 16, France, or to OECD Publications and Information Centre or Distributor in your country *(see last page of the book for addresses)*.

Prices charged at the OECD Bookshop.

THE OECD CATALOGUE OF PUBLICATIONS and supplements will be sent free of charge on request addressed either to OECD Publications Service, or to the OECD Distributor in your country.

CONTENTS

ACKNOWLEDGEMENTS

The author wishes to acknowledge the contributions and helpful comments of Mario Amendola, Pascal Petit, Gabriel Tahar and Charles Oman, and thanks Julie Gaskill for editing the manuscript.

The Development Centre expresses its gratitude to the Government of Norway for its support to the Centre's research on New Technologies and Changing Comparative Advantages in Industry, of which this study is a part.

PREFACE

Few changes are having a greater impact on the ability of firms and countries to compete in global markets than the revolution in telecommunications. The new capabilities of information processing and transmission are profoundly transforming requirements for human skills, for capital equipment and for corporate strategies in countless manufacturing and service industries. This transformation is affecting both developed and developing countries.

Indeed, as this study points out, the new technologies are rendering obsolete an important part of the existing telecommunications equipment and infrastructure in many of the most advanced industrialised countries. The result of this technological upheaval is to offer important opportunities to late investors – to late industrialising countries – for catching up and even "leapfrogging" the industrialised countries that invested more heavily in telecommunications during the 1960s and 1970s.

To the surprise of many, the effective diffusion of the intertwined system of information and communications technologies has been especially fast not in the OECD countries where the new technologies were first developed, but in countries with low levels of pre-existing communications infrastructure and low levels of costs sunk in the telecommunications system. This unexpected pattern of diffusion reflects the fact that it is easier to diffuse systemic innovations from scratch than to add to existing systems on a piecemeal basis.

Such leapfrogging possibilities have also been favoured by the fierce global rivalry among the handful of OECD-based multinational corporations that supply telecommunications equipment, particularly switching equipment, worldwide. This competition is reflected in the plummeting of the prices of the new electronic switching equipment, first introduced in the United States in the early 1970s, by a factor of three, in current dollars, between the mid-1970s and the mid-1980s, thereby providing a major stimulus to the global diffusion of the new technology.

The result, moreover, has been to open new opportunities for some developing countries and especially NIEs to close the productivity gap *vis-à-vis* OECD countries. The international competitiveness of Filipino and Taiwanese textile, garment, toy and consumer-electronics producers, for example, has grown significantly in recent years through their increasing use of advanced telecommunications services to forge closer links with distributors in North America and Europe. The fast-growing reliance of Korean firms on data networks to co-ordinate flows of components and intermediate goods as well as international marketing and financial information, and the extraordinary efficiency of financial markets in Singapore and Hong Kong, which play an important role in Pacific Asia as a whole, are further examples of the increasing use of advanced telecommunications services by NIEs to strengthen their competitiveness in the global economy.

The present study, written by a leading telecommunications economist, provides an in-depth analysis of the ways in which developing countries can overcome information asymmetries and other imbalances to take advantage of opportunities for catching up and leapfrogging. In doing so it provides policy makers seeking to enhance the global competitive strength of their countries with a ground-breaking, comprehensive view of the importance of the new telecommunications technologies. It thus represents a key contribution to the debate on the two critical questions for the years ahead: Will developing countries gain access to the new technologies? And, in so far as they do, what are the implications for global trade and competition, for OECD as well as for developing countries?

Louis Emmerij
President
OECD Development Centre
April 1991

EXECUTIVE SUMMARY

Electronic switching, introduced in the United States in the early 1970s and a key component of advanced telecommunications, has been quickly adopted by many countries since the late 1970s. Most of the fast adopters have been newly industrialising countries, which have shown an extraordinary capability for acquiring and using the new technology. Many industrialised countries that have been powerful contributors to technological progress in telecommunications, on the other hand, have lagged behind in its diffusion.

This uneven and unexpected pattern of technology diffusion deserves attention for several reasons.

First, economic historians have often reminded us of the delays involved in the diffusion process. For a significant technological innovation to reach most potential users a wait of 40 years or more is common. This is partly because diffusion has always been expected to occur first in countries with economic features similar to those of the innovating ones.

Second, electronic switching and advanced telecommunications are at the core of the emerging information technology. A country's ability to benefit from new information systems, through domestic and international computer networks, data communications, fax and media, for example, depends heavily on the modernity of its telecommunications network. Adoption of electronic switching is therefore a prerequisite for the diffusion of other information technologies.

Information, comprising both product and organisational innovations (a firm adopting innovated goods often changes its structure and routine) and, more generally, knowledge, is an "imperfect" commodity. Its costs do not always reflect concrete, measurable profits to the buyer. Creating, using and transferring knowledge depends on dynamic externalities such as the reduction in price of innovated goods brought about by increases in sales, economies of scale and learning by doing; economies of learning in the use of innovated goods; reduced prices of complementary factors like skilled manpower, maintenance, spare parts, software programmes, etc., and introduction of incremental innovations by upstream industries. Countries and firms that can acquire the necessary information concerning innovations and then integrate them usefully into their industrial structures are those most likely to experience high rates of growth.

Awareness of the speed and direction of the international innovation of technological diffusion, its determinants and its limits is important for assessing growth opportunities for developing countries. These countries may have a poor capacity for generating innovations, but they are not excluded from adopting them quickly and effectively. In fact, the diffusion of advanced telecommunications has been especially fast to some less advanced countries.

The study of international technology distribution patterns has been heavily influenced in the past two decades by the product-cycle approach of the 1960s. This approach suggests that adoption of innovations by less industrialised countries depends upon those countries' financial and technological capacities, and that a precise succession of events invariably takes place:

1. Advanced industrialised countries introduce technological innovations that are diffused first within their domestic markets;
2. When users in less developed countries learn of the new technology's profitability, they import it, often at a relatively high cost;
3. As the new technology diffuses within and among the advanced countries its purchase price declines, and the establishment in less developed countries of both local "imitating" firms and subsidiaries of multinational firms from the more advanced countries leads to some production of the new technology as well as growing local technological capability in the less advanced countries.

In this view, exposure to international trade and to the foreign direct investment that follows exports from the innovating countries are necessary for less developed countries to be able to imitate the innovating ones.

The product-cycle approach has traditionally focused on the purchase price of the capital goods that incorporate the innovation, their operating costs and profit potential, and the economic features of the adopter population: size and distribution of firms, wage and interest rates. We can label this rather narrow view the "microeconomic" approach to explaining international technology diffusion.

In the 1980s a radical change in thinking evolved that analysed innovation diffusion more globally. This new approach allowed for the possibility of leapfrogging; i.e., for the possibility that late industrialising countries can assimilate technological innovations more quickly than earlier industrialised ones, and that diffusion capacity can overtake innovation capacity, causing innovative countries to be laggards in adopting.

This recent "macroeconomic" view considers investment flows and the growing rate of demand for the innovation, as well as the innovation's specific technical features as primary factors in international diffusion. According to this approach, the factors that explain the diffusion of advanced telecommunications technology can only be assessed globally because of its strong "system" characteristics: it requires a strict network of technical complementaries and interconnections between components to be successfully adopted.

Because of advanced telecommunications' network characteristics, its diffusion calls for the complete replacement of any pre-existing telecommunications infrastructure. This is in line with Frankel's observation years ago that, "As an industry grows and adapts to changing and increasingly complex production methods, interconnections, more or less rigid, develop among its components – among machines, plants, transport networks and raw material supplies – that make increasingly difficult the introduction into the system of new, cost-saving changes.... As a result, [its] costs are higher and labour productivity lower than it would be in a less 'mature' industry. The old industry finds itself penalised for having taken the lead and shown the way to its young competitors in other regions[1]."

The effect of this technical exigence of advanced telecommunications on its international diffusion has rarely been appreciated. It has three important consequences:

1. It is likely to be faster in countries with high growth of demand, because of the strong expansion potential of telecommunications lines;

10

2. It is likely to be faster in countries with low telecommunications density, less "sunk" costs in a telecommunications infrastructure at the time of the innovation's introduction; and
3. It is likely to be faster in countries with high investment potential in telecommunications.

Advanced telecommunications, then, seem to offer the perfect example of a major leapfrogging opportunity. Fast adopters tend to be countries that have not already invested heavily in an existing telecommunications network and so are technically and financially more open to new technologies. Because these countries have not committed themselves to a structure of supplies, equipment, plants, etc., that would have to be scrapped before innovations in telecommunications could be implemented, and because of an ongoing decline in the innovation's purchase price, late-comer countries have lower start-up costs. Their labour is less expensive than that of more industrialised countries, and they are more exposed to diffusion of information through international trade. They are more receptive to imports, with less need to support local suppliers because their markets are smaller.

Latecomers with burgeoning economies can also benefit from a high investment rate. Risks associated with the irreversibility of investment are more easily run by imitating countries than by those that have already invested intensively in capital goods.

The speed of diffusion among adopters varies considerably, of course; it depends on the investment climate, interest rates, profitability comparisons between the new capital goods and existing ones, the performance of the technology in other adopting countries, the inefficiency or even non-existence of the adopting country's technological infrastructure, availability of technical assistance and skilled manpower, wage rates in the adopting country (generally the higher wages are, the faster the adoption), and research and purchase costs. Late adopters usually decide to acquire the innovation when the entire technological system has been completed and assessed. They can then start from scratch without worrying about the depreciation of an older technical system.

Once the diffusion of advanced telecommunications is under way, its benefits can spread throughout the entire economic system. By paving the way for the diffusion of modern information technologies, it can be the key to boosting a country's overall growth rate and competitiveness on world markets. To achieve continuous economic growth, a country's readiness to adopt new processes and innovations is as important, if not more important, than the capacity to innovate them.

The strongest opportunities for catching up and fast diffusion of advanced telecommunications seem to exist for countries that:

a) Have a small economy;
b) Are exposed to high levels of international trade and foreign direct investment;
c) Experienced high demand for the new technology at the end of the 1970s, when electronic switching was well established (instead of, for example, in the 1960s, when high demand would have warranted investment in electromechanical lines);
d) Have low levels of pre-existing telecommunications infrastructure;
e) Are free to reshape "vertical relations" between local suppliers of telecommunications equipment and buyers (telecommunications carriers who purchase digital switching) instead of depending on supply from established affiliates of a foreign firm.

Substantial opportunities for catching up have also been created by a specific combination of changes on the supply side:

a) The introduction of digital switching obliged telecommunications equipment manufacturers to spend large amounts on Research and Development. This increased not only their "minimum efficient size" – they needed to expand their markets to recoup these costs – but also the height of entry barriers to new competitors, and consequently the concentration of supply.

b) The transformation of electromechanical switching to electronic and eventually digital switching toppled the vertical integration that had existed between telecommunications equipment manufacturers (affiliates of multinational firms) and local customers. At the same time it lowered mobility barriers to manufacturers who had traditionally been tied to their branch markets, but who could now sell internationally.

c) The globalisation of markets for digital switching equipment intensified competition among the few manufacturers. Apart from reducing their average costs in R&D by acquiring new markets, manufacturers could also take advantage of "short-term dumping strategies" – in other words, selling the first units of a complete (and basically non-interfacing) infrastructure at a greatly reduced price and thus "locking in" long-term customer loyalty.

d) The telecommunications carriers' monopolistic, heavily regulated structure in each country gave them bargaining power *vis-à-vis* manufacturers, at least in the short term. Once they were locked in to a manufacturer, of course, they had little choice but to continue with whatever system they had begun.

These conditions and the sequence of events that produces them appear rather specific and perhaps unlikely to be duplicated in other industries and technologies. Still, with this in mind, this study provides an in-depth analysis of the ways in which developing countries can overcome information asymmetries and other imbalances to take advantage of catching up and leapfrogging opportunities.

Chapter 1 reviews the major debates on the causes and patterns of international innovation diffusion. It pays special attention to the swing from the product-cycle approach to the more recent macroeconomic emphasis on economic growth and the innovation's technical aspects as determining factors in its diffusion, and how these apply to leapfrogging and catching up opportunities for less advanced countries.

Chapter 2 focuses on "interrelatedness"; i.e., the "system" aspect of advanced telecommunications, and how this affects its international diffusion. Special attention is paid to the effects of adoption on the productivity and competitiveness of downstream industries and of the overall economy.

Chapter 3 examines the empirical evidence on the international diffusion of the core component of advanced telecommunications: electronic switching. This chapter will demonstrate that diffusion rates were particularly high from 1977 to 1987, especially compared to other key technological innovations of recent years; that diffusion during this period was not only remarkably fast in Far East Asia, and to a lesser extent, Latin America, but it was faster in most late industrialising countries than in many early industrialised European countries.

Chapter 4 analyses the role of supply, market structure and industrial organisation in the international diffusion of advanced telecommunications. The introduction and application of microelectronics has shaken the international telecommunications equipment industry, completely altering supply and demand. High R&D costs have increased the slope of long-term average cost curves and raised barriers to new competitors, while improving mobility for existing manufacturers, and technological discontinuity between electromechanical and electronic switching has erased the advantages of vertical integration, especially in small countries. Consequently the number of captive markets has fallen and competition has intensified.

Chapter 5 tests the importance of macroeconomic conditions, and more specifically investment, in the adoption of technological innovations. It shows that the catalyst for the unusually fast rates of diffusion of late industrialising countries has been investment. Countries possessing both good adoption capacity due to strong cognitive and real externalities (like learning opportunities and expertise), and large-sized firms that provide telecommunications services, have stimulated investment in innovated equipment and experienced rapid modernisation.

Chapter 6 then presents general conclusions on the international diffusion of innovations and policy implications for developing countries.

NOTE

1. M. Frankel, "Obsolescence and Technological Change in a Maturing Economy", *American Economic Review,* (September 1955), pp. 296-319.

Chapter 1

ECONOMIC GROWTH AND THE INTERNATIONAL DIFFUSION OF INNOVATIONS

A revival of interest in the influence of knowledge access, information and technological change on international technology diffusion has occurred in recent years. The classical view that increasing returns are cumulatively generated by the rates of growth of output and investment has been strengthened by the growing belief that information availability and other externalities play a vital role in such a process. It is the concept of information and knowledge as a highly imperfect commodity (in that it cannot be measured or priced) that has shaped current thinking on development, and has led to the following propositions:

1. Information and knowledge, like all public goods, rarely stay for long in the exclusive possession of the firms that produce them[1];
2. The value of information is difficult to assess and therefore not easy to trade[2];
3. Because knowledge and information are difficult to transfer on markets, customers have a hard time acquiring them[3];
4. The production and acquisition of information are themselves the joint result of learning processes[4];
5. Learning (and "learning to learn") is a cumulative process that generates considerable increasing returns[5];
6. Market failures will probably occur because of the monopolistic strategies of incumbents and early market entrants[6];
7. As the generation, capture and handling of knowledge become increasingly important, and as it is such an imperfect commodity, organisations will eventually supplant markets, and co-ordination will replace market transactions[7].

According to many economic analysts, information asymmetries and externalities provide a dependable gauge for understanding a country's productivity growth as well as international differences in overall growth rates and in the speed of generation, adoption and diffusion of technological change across countries.

Romer[8] has made a model of long-term growth in which knowledge figures as an intermediary input, as essential to productivity as data to a computer. When a firm makes an investment in knowledge, it creates an additional stock of knowledge, the sum of which is greater than its parts, and ultimately all firms benefit. As a matter of fact, according to Scott[9], all investment increases knowledge and experience, and produces a higher rate of return for the public than for the private sector.

Already Nelson had pointed out that access to information is a decisive factor in a country's improved productivity and, more generally, in its overall economic growth: "Viewing the economic development process as a diffusion process naturally leads one to abandon the two basic assumptions of the neoclassical model – that all firms in all countries are on the same production function, and that markets are all in full equilibrium"[10].

In view of this it seems necessary to discover what propels international diffusion, specifically in the case of advanced telecommunications, and what the implications are both for economic strategies and for policy modelling. We will start by briefly reviewing the current debate on international diffusion.

Two often-conflicting views are juxtaposed; the former, elaborated in the mid-1960s, is based on a microeconomic framework of analysis built upon the product-cycle approach and the epidemic modelling. It links a country's capacity to adopt with the accepted routines of imitative adoption and the international differences in relative factor costs and price of the new capital goods. The latter, developed in the mid-1980s and still evolving, focuses on the macroeconomic conditions of adoption such as investment and demand, and on the technological characteristics of the innovation.

To understand the causes of international differences in speed and intensity of technological innovation diffusion we should consider both *demand* and *imitation lags*. Demand lag is the time that elapses before import and use of final products and capital goods such as, in the case of advanced telecommunications, electronic switching hardware and digital transmission systems. Imitation lag, on the other hand, is the time it takes a country to begin to manufacture the innovation and create new services such as, again in our case, data communications. In advanced telecommunications demand and imitation lags closely parallel each other. The adoption of digital switching enables telecommunications carriers to supply advanced services that incorporate technological innovations (facsimile machines, for example) and that consequently induce potentially profitable organisational innovations within the manufacturing and service sectors.

In a product-cycle approach the speed and extent of international diffusion of innovations is expected, first, to depend on each adopting country's wealth and technological capacity, and second, to follow the pattern of domestic diffusion within advanced countries.

This approach leaves little room for the hypothesis developed in the 1980s of leapfrogging, i.e. that late-industrialising countries are capable of adopting technological innovations faster than early industrialised ones, and that adoption capacity can differ enormously from innovation capacity, so that innovating countries can be laggards at adopting.

The Microeconomics of the International Diffusion of Innovations

Throughout the 1970s and most of the 1980s the debate on the determinants and patterns of international innovation diffusion was deeply influenced by the product-cycle approach developed by Vernon[11] and Hirsch[12]. Their analysis proposed that a country's revenue levels and technological capacity, openness to trade, foreign direct investment and ability to diffuse technology are intertwined, and that a precise sequence of events is expected to take place:

1. First, advanced industrialised "producers" with high levels of technological capacity engender innovations that are diffused only within their own domestic market;
2. Second, these innovations are diffused abroad only when "users" in less advanced countries learn about the profitability of adopting the new technologies;

16

3. Less advanced countries import the new technologies from the more advanced countries, usually at relatively high prices;
4. Finally, as wages increase and purchase costs diminish, the adopting country's capacity to manufacture the innovation on its own, as well as its growing technological ability, improves. The entry of multinational firms from the innovating countries and imitation by local companies also helps diffusion of production and technology capacity.

The scientific and technical ability to create new technologies and the incremental innovations that expand their scope and profitability is concentrated in a few advanced countries. These countries benefit from strong domestic demand that favours the introduction of new technologies, already possess a high level of technological sophistication and have the high wages that permit widespread consumption of innovations. This is definitely the case in telecommunications in the United States, where an advanced electronics industry grew out of insistent demand from the commercial, financial and manufacturing sectors for more efficient information handling.

In a product-cycle approach, after such an apparently successful innovation spreads domestically, international diffusion among users starts. Adopting countries with healthier economies and more pronounced similarities to the innovating country will take less time to import the new technology; in other words they will experience a shorter demand lag.

In the product-cycle approach, then, demand lag is necessarily shorter than imitation lag, and always occurs first. New products, in fact, are usually obtained in response to domestic demand. In its first stages the demand of adopting countries is satisfied by exports from the innovating countries. Eventually, though, production starts and they become imitators.

The stronger the domestic demand for a new technology and the more important the incremental innovations that improve its usefulness, and the better the learning opportunities and advantages from economies of scale (larger plants can mean lower production costs) then the shorter the time for adopting countries to manufacture the innovation and create services.

The debate on international diffusion has long been influenced by the contributions of Posner[13] and Vernon[14]. Posner sees imitation lag as the result of three developments:

 a) A foreign reaction lag;
 b) A domestic reaction lag;
 c) A learning period.

A technological innovation is usually acquired by foreign markets through exports from the country where the innovations were first engendered. The foreign reaction lag is the time it takes the first importer to imitate the production of the innovated goods. In a product-cycle approach, the length of the foreign reaction lag depends on the intensity of the threat to domestic markets posed by these foreign imports.

The domestic reaction lag is the time that evolves before local producers supply a relevant share of the domestic market. The extent of this lag is expected to depend on the size of the internal market; the larger it is, the shorter the lag.

The amount of time necessary for an innovation to be understood and successfully applied (the learning period) differs among countries for many reasons. Posner concentrates on the volume of total past output as an indicator of technical competence. Kaldor[15] feels that the length of time a country has been manufacturing a product is a more appropriate measure of the learning period.

Swann[16] has studied the reasons behind international differences in international rates and levels of diffusion of synthetic rubber in North and South America, Europe and Southeast Asia between 1953 and 1969. According to his findings, variations in the origins of diffusion – the date at which an innovation reaches penetration levels of 10 per cent (hence, in his particular study, when synthetic rubber made up 10 per cent of total rubber consumption) – were determined by the amount of imports per capita from the United States (the innovating country), the volume of local production of tyres per capita, and the amount of time each country had been producing synthetic rubber. Swann found that diffusion was faster a) when the origin of diffusion was later; b) when the growth rate of the rubber goods industry was faster; and c) the greater the imports of rubber goods per capita. Swann's study is clearly influenced by the product-cycle approach: diffusion among users predates diffusion among imitators; i.e. demand lags are shorter than imitation lags.

The results of the empirical analysis by Tilton[17] on international diffusion of semiconductor technology reveal the importance of market structure. In the United States, the innovating country, rapid imitation of the technology was undertaken by many small new US firms. The situation in imitating countries, however, was quite different. In Europe diffusion was assured mainly by subsidiaries of US multinational companies. In Japan production was initiated by large diversified groups that already used semiconductors and were big former clients of US firms.

The European and especially the Japanese experience seem to indicate that large established firms are a requirement for an imitating country to catch up. Ozawa's study[18] of Japan's acquisition of synthetic fibre technology further demonstrates the importance of co-operation between government policy and industry. According to the "staggered-entry" formula used in a government programme only one firm from each Japanese industry chosen was allowed to enter into international bargaining to obtain relevant technology from abroad. To qualify as an entrant firms had to show a real command of the innovated technology, and to do this they had to conduct a lot of "backward engineering" – studying one component of the innovated technology to try to understand the whole – and "patent-literature-based reproduction" – reading patent literature for information about the product. Obviously only large firms could mobilise the necessary resources and fund the investments necessary for adoption. As a consequence, however, diffusion of new technologies spread very fast. This was partly due to the enha .ced bargaining power of Japanese firms, which were inundated with foreign suppliers.

Stobaugh[19] shows that the major determinants of imitation lag in nine product innovations in the chemical industry were, in order of importance, market size, availability of local technology, the investment climate and shipping costs. Market size affected the order in which imitating countries began manufacturing the innovation: countries with small markets naturally began the manufacture of technological innovations later. Countries that preferred to import rather than protect local development of the technology (an expensive undertaking because of high manufacturing costs and learning difficulties) tended to initiate diffusion about ten years earlier. A country's investment climate (the result of inflation rates, currency stability, import and export tariffs and general expectations about political and economic change) is the third major factor in judging the duration of a country's imitation lag, and Stobaugh's findings indicate that countries with a good investment climate began production approximately ten years before countries with a depressed investment climate.

The main determinants of the delay that precedes import and use of an innovation appear to be information asymmetries and learning times. Not all potential adopters are informed at the same time about a technological innovation's characteristics or the profitability of its adoption. Learning times depend on the potential adopters' opportunities to evaluate the new technology, especially by assessing the stock of innovations of early adopters.

The decision to adopt (or not) based on an awareness of what others are doing – the classic *epidemic* process – is therefore a key element in diffusion, and accounts for international differences in the timing of adoption of new products and processes. The epidemic methodology developed by Mansfield[20] and Griliches[21] has laid the foundation for most empirical analysis in this field.

Firms rely on each other to learn about a new technology and ways of using it. Basically we can conclude that 1) it is costly for all firms to learn about and use new technologies[22], 2) firms differ in their ability to learn about new technologies and 3) learning externalities such as learning times and accessibility of skilled manpower significantly affect decisions about adopting[23].

Recent developments have highlighted the strategic role played by network externalities – signals ranging from market conditions to costs or profitability of other firms using the technology – in understanding the environment required for the adoption of innovations and the new capital goods that interlock with them. The effects of these externalities are in fact particularly evident in innovations such as telecommunications-based services and equipment, where a notion of the *cost of adoption* has emerged as distinct from the usual *cost of purchasing* (the market price of new capital goods). The profitability of new capital goods in the light of externalities depends only to a limited extent on the prices of the goods on the market. Much more relevant are the costs of adoption and use. These costs stem from the behaviour of other firms (which technology they adopt) and on the penetration of the technology in the region. As the number of adopters increases, so does the availability of skilled manpower; maintenance and spare parts costs decline and collective learning improves (Katz-Shapiro 1985).

Empirical evidence suggests that the rate of diffusion of new technologies across countries is affected by the strong cumulative effects of "interrelatedness"; latecomers (late-adopting countries) with lower levels of penetration – smaller amounts of the innovative capital goods that embody the new technology – may also have lower rates of diffusion[24]. This is because it is difficult to obtain all the relevant information and skills for information-based technologies without having already reached a certain level of interrelatedness with other complementary technologies[25].

Much progress has been made because of David's historic analysis of the diffusion of the QWERTY typewriter keyboard[26]. David was the first to realise the powerful effects of network externalities, i.e. the intertwining of consumption, pecuniary and technical externalities which shape diffusion processes. Network externalities are caused by the reductions in the cost of training, maintenance and spare parts that ensue from an abundance of users, as well as by the reductions in installation costs engendered by increases in the scale of production of complementary equipment and supplies. More recently, David and Bunn described the notion of "network" technologies, which "are characterized by some form of increasing returns in use or production and ... entail choices where considerations of technical interrelatedness among the components forming alternative systems matter[27]".

Analysis of the distinct network features of information technologies requires an understanding of the role played by externalities in general in the adoption of capital goods that incorporate innovations. Careful examination of the pattern of adoption and diffusion of a wide variety of new technologies suggests in fact that "network externalities", or more generally "dynamic externalities", apply to many organisational innovations as well. Dynamic externalities are characterised not only by technical interrelatedness and vertical complementarity in upstream and downstream markets; in the case of advanced telecommunications, diffusion of digital switching and the transmission of data, it enables manufacturers to fashion efficient organisational innovations such as "just-in-time" manufacturing techniques that cut down on warehousing and related costs. Dynamic externalities appear to be especially strong at the regional level because of this tight meshing of product and process. In some, within a region, the cost of adoption and use, as distinct from the cost of purchasing, is influenced by:

- the availability of information about the technology from other users;
- the availability of trained skilled manpower;
- the availability of technical assistance and maintenance;
- the availability of complementary equipment and software;
- the availability of complementary innovations, both technological and organisational.

When these resources are abundant, the costs of adoption are expected to be lower and – for given costs of purchasing – profitability will be higher, with consequently higher diffusion levels.

Moreover, as information spreads and the stock of innovated capital goods grows, transaction costs decline; therefore imitation takes place through learning from others and collective learning.

In sum, according to the epidemic and network externality approach, information asymmetries and learning times provide the basic explanation for differences in the origin of diffusion, and diffusion rates are mostly explained:

a) By encouraging signals (besides prices) that motivate firms to overcome information asymmetries and acquire the essential knowledge for the adoption of the innovation;

b) By the presence of "advanced users" who help to push the diffusion through their introduction of innovations that improve the new technology[28].

International differences in the rates of diffusion of new technologies clearly are determined by the dates of origin of the diffusion process and by the timing of adoption by advanced user countries. Adoption by these users will very probably activate the epidemic process of learning through others, overcome the initial attrition of an early lack of information and stimulate a phase of rapid growth in the stocks of adoption.

In an analysis of international diffusion of the oxygen steel-making process from 1956 to 1964, Maddala and Knight present some reasons for international differences in innovation adoption rates[29]:

1. Differences in the relative prices of labour and capital among countries;
2. Different economies of scale between processes (large factories with extensive equipment can produce more cheaply);

3. Differences in the type and quality of intermediary inputs such as hardware and machines available in the adopting country;
4. Differences in product flexibility between processes (the variety of products that can be manufactured with existing machinery);
5. Differences in the age of the industry's existing capital goods;
6. Differences in the rate of growth of the industry;
7. Differences in exposure to international trade.

The results of Maddala and Knight's empirical analysis focus especially on the role of exposure to international trade. Those countries lagging in the adoption of the basic oxygen furnace technology were also those with the highest barriers to international trade.

Chen studied the diffusion of 45 technological innovations (all originated abroad) within five industries in Hong Kong: textiles, garments, electronics, plastics and toys[30]. In these cases profitability of adoption appeared to have the strongest influence on diffusion rates. Large unitary levels of initial investment for adoption tended to delay diffusion in the electronic sector and in textiles. Affiliates of multinational firms which had access to information about the innovation from their headquarters helped diffusion. In addition, empirical results also confirm that high levels of concentration are conducive to fast diffusion rates. Chen's conclusions closely parallel those of Tilton and confirm that multinational and large firms are necessary for diffusion to take place in imitating countries.

The importance of investments in the adoption of a new technology demonstrated by Chen's study corresponds to results obtained by Hsia, who found profitability and amounts of initial investment to be the major determinants of diffusion rates of 26 technological innovations in Hong Kong in the plastics, textile, garment, electronics and electrical engineering industries[31].

Stoneman places much emphasis on the role of wages in the adoption of new technologies[32]. In his opinion early adopters are always countries in which wages are high. This is a natural conclusion of the "equilibrium" approach that considers the time spread between early and late adopters as primarily the result of differences in profitability of adoption – profitability that is gauged by the costs of production factors (including, of course, wages).

In this scenario, then, late adopters are not agents affected by information asymmetries but rational investors who prefer to continue using older techniques if the innovated ones are capital intensive, at least until that time when wage increases cost as much as adopting the new technology. Consistent, then, with the equilibrium approach is the significance of the price of innovated capital goods for adoption.

Nabseth and Ray's extensive review of the diffusion rates of ten technologies among and within six European countries also concludes that wage levels are the most decisive factor when profitability of adoption is assessed. Next were the quality of information available and the pre-existing inter-industrial linkages[33]. In their study, countries in which local supply of intermediary or complementary inputs was scarce and of poor quality, and industrial demand for the product was high, adopted faster. These inter-industrial linkages were vital, for example, in the rapid diffusion of numerically controlled machine tools because of the tools' usefulness in the aerospace industry, of shuttleless looms because of the innovation of man-made fibres, of basic oxygen furnaces owing to the lack of scrap, and of energy-saving tunnel kilns because of high energy costs.

One of the most significant results of Nabseth and Ray's analysis is the importance of forward and backward linkages, i.e., the supply and demand of technology-intensive products throughout industry, in analysing international differences in innovation diffusion. In a more domestic framework David[34] highlighted the role of inter-industrial linkages between agriculture and industry in explaining the rapid diffusion of technological innovations in the area surrounding Chicago. Similarly Nelson, Peck and Kalachek[35] stressed the incentives to adopt that appeared once bottlenecks such as shortages of supply and manufacturing capacity were removed and prices were subsequently allowed to decline.

Lynn's study of the diffusion rates of the basic oxygen furnace (B.O.F.), based on an estimate of the growing capacity of US and Japanese steel plants, shows that of the two countries Japan, because of its much higher scrap prices and growth of demand, adopted B.O.F. faster[36]. The limited supply of scrap in Japan had hampered the growth of the steel industry, and Japanese steel producers thus had a powerful incentive to adopt a technology that required lower quantities of scrap than open hearth methods.

A recent study examined the determinants of international diffusion of information and communication technologies based on penetration levels and adoption rates of modems (devices that convert data to a form that can be transmitted via telephone) in 16 countries[37]. The results of the study showed that the origin of diffusion, the time necessary to reach 10 per cent of total potential demand, was earlier the smaller the size of the country, the larger the levels of revenue per capita, the smaller the intensity of R&D and the larger the ratio of employment in US multinational firms to total employment. The speed of diffusion was higher in countries where diffusion started earlier and where the telecommunications infrastructure was relatively advanced and sophisticated.

These results confirm that small countries can benefit from new technologies earlier than large ones for two reasons: 1) they are more exposed to international trade and therefore the diffusion of information and 2) they are more open to imports and have less need to protect local suppliers because their markets are small. This is especially true when their markets are too small for minimum efficient size of production of the innovated goods.

In another recent study, on the speed of diffusion of shuttleless looms between 1976 and 1985 in nine countries, it was established that an innovation's adoption within a country is influenced not only by network externalities and the imitation processes but by the lowering of the innovation's purchase price on international markets[38].

In conclusion, the analysis of international innovation diffusion in the microeconomic approach can be characterised as follows:

– Its focus is the importance of information asymmetries on the firms' decision-making;
– International diffusion is expected to take place in imitating countries only after domestic diffusion within the innovating country;
– Rates of international diffusion depend on factor costs – labour and capital – and the presence of large firms and affiliates of multinational firms in imitating countries, as well as on the forward and backward linkages of old and new technologies;
– Exposure to international trade and foreign direct investment are essential to a country's ability to quickly imitate advanced ones. All limits to imports delay diffusion by reducing information flow. Similarly, all custom burdens on imports delay diffusion because they reduce adoption profitability.

In a product-cycle approach the polarisation of the two groups of countries – innovating ones and imitating ones – is obvious and no change in their positions seems likely: domestic diffusion within the innovator will always be faster than among imitators.

From this point of view, multinational firms can help imitating countries adopt quickly, but foreign direct investment that follows on the heels of exports from the innovating country is an absolute requirement.

The Macroeconomic Context of Adoption: Catching Up and Leapfrogging

The debate on the international diffusion of innovations has undergone a transformation since the appearance in the 1980s of a "macroeconomic" theory of diffusion. Japan and Italy's example of catching up with and leapfrogging older industrialised countries like the United Kingdom has pushed many economists to re-examine traditional views of economic growth and technology diffusion.

Growing empirical evidence shows that for many new technologies diffusion in fast-growing imitating countries has been quicker than in innovating ones.

Key variables in measuring international diffusion are economic growth rates (and more specifically, rates of growth of demand) and investment. A pathbreaking study by Soete called attention to the role of investment and the age structure of existing capital goods[39]. In his perspective, late-industrialising countries with low amounts of past investment and high levels of present investment have a good chance to leapfrog technological leaders, both users and producers, to adopt innovations. As Soete recalls, such a process was at the heart of the rapid expansion of the Japanese and Italian economies whose eventual leadership in the sectors of steel, automobiles, electronics and numerically controlled machine tools grew chiefly from imported technology.

According to Abramovitz, in leading countries the "technological age" of the capital stock, defined as the amount of technological innovations the stock contains, is the same as its "chronological age" (real age)[40]. In follower countries, where total productivity is lower, the technological age of the stock is higher than its chronological one. The profitability of adopting an innovation (after discounting for amortization costs) is higher in follower countries than in innovating ones. Investment opportunities (and consequently adoption) are therefore higher in follower countries.

Abramovitz' line of thinking leads to two main hypotheses:

1. "Backwardness offers an opportunity for modernisation in 'disembodied' technology (in the form of patents or licenses) as well as 'embodied' technology (the knowledge embodied in capital goods);
2. Rapid progress in modernisation encourages fast growth of capital stock. So besides a reduction in technological age towards chronological age, chronological age also goes down. The rise of the ratio of capital-labour speeds up. Productivity benefits on both counts."[41]

Catching up can lead to leapfrogging, particularly when the new capital goods incorporating the technological innovation cannot be added piecemeal to existing ones, but require the scrapping of all pre-existing capital stock, regardless of its age structure. This implies that imitating countries can adopt innovations earlier and faster than innovating countries, owing to the better investment opportunities and profit potential inherent in the young age of their capital stock. Consequently leapfrogging possibilities are especially important for countries that start out with insignificant amounts of capital goods and are obliged to "start from scratch".

The empirical evidence of these catching up and leapfrogging possibilities is abundant. Ray's founding contributions[42] are some of the richest and most elaborated analyses of the patterns and causes of international innovation diffusion. Ray spent almost 20 years studying the diffusion of oxygen steel-making, continuous casting of steel, tunnel kilns in brickmaking, shuttleless looms, float glass and numerically controlled machine tools. Here are some of his conclusions:

1. The magnitude of the indivisible investment necessary to buy a new technology and the technology's "discontinuity" (lack of compatibility with pre-existing technologies) are major causes of delay in adoption. Technological innovations that can be adopted piecemeal within the existing industry diffuse faster.
2. To successfully evaluate diffusion rates, the macroeconomic environment of adoption must be considered. "In boom times, when industries and companies expand, when capital is easily obtainable and investment is high, conditions for the quick spread of new technology are favourable. Years of stagnation or slump, on the other hand, lead to austerity programmes with reduced investment activity which could put a brake on the dissemination of new technology as well[43]".
3. Profitability of adoption depends on a variety of factors: the market price of the innovated capital goods, wage levels, access to capital and its cost, costs of complementary inputs such as energy and raw material and potential prices of final products and services.

In an analysis of the surprisingly slow diffusion of process innovations in the United Kingdom, a country that is traditionally a robust technological leader, Ray's findings confirm that the age of the capital stock, the macroeconomic context of adoption (rates of economic growth and capital accumulation), the technological "interrelatedness" of the innovations and standard microeconomic factors like wages, interest rates, size of firms and information asymmetries explain why the United Kingdom has not kept pace with its competitors.

Another analysis of the contrast between slow rates of domestic diffusion in innovating countries and fast international diffusion among imitating ones is offered by the case of open-end spinning rotors[44]. A study of the effects of technological expectations in this industry, i.e. expectations about the eventual introduction of incremental innovations, on the diffusion of open-end rotors in 26 countries between 1975 and 1985 suggests that the gap between innovation capacity and diffusion capacity widens when incremental innovations are expected. As innovating countries become aware of them, they tend to delay adoption because of the risks associated with irreversibility of investment.

Diffusion has been described as the consequence of investment behaviour and adoption ability, influenced both by macroeconomic factors like demand growth and the traditional epidemic learning process of judging what advantages others are gleaning from the new technology[45]. An empirical analysis of the diffusion of shuttleless looms in 16 countries during the period 1977-1984 shows that diffusion is indeed the combined result of a microeconomic process of learning and a dynamic macroeconomic context of adoption conducive to healthy investment.

Part of the renewed interest in the relationship between economic growth and international innovation diffusion since the product-cycle approach of the mid-1960s has centred on the powerful effects of inter-sectorial flows of technology. In other words, international diffusion of process innovations benefits not only the adopting industry but "downstream" industries. Fast diffusion in the "upstream" sectors of an imitating country will improve overall productivity and competitiveness, increase downstream industries' technological capacity and proclivity to adopt other innovations, and eventually produce a snowballing effect that can threaten competing innovative countries.

The textile industry once again provides ample evidence of such a process. The rates of adoption of shuttleless looms by 28 countries from the mid-1970s to the mid-1980s have had a striking effect on international competitiveness in the cotton fabric markets[46]. Countries that were able to adopt and use the looms faster, like Hong Kong, Thailand and many less advanced OECD Member countries, had higher market shares. This indicates that:

1. Technological advancement is better measured by diffusion rates than by R&D expenditure in user countries;
2. Diffusion exerts a strong effect well beyond each industry.

Dalia arrives at similar conclusions[47]. She describes how the Austrian textile industry achieved strong productivity through imports of innovated textile machinery that could not be produced locally. Petit, Tahar and Antonelli confirm that the availability of complementary innovations in upstream industries influences the rates of diffusion within the textile industry. Backward and forward linkages appear to matter not only in the adoption of isolated innovations but also in broader technological systems that use a wide range of innovations. The diffusion of open-end rotors in 16 countries from 1976 to 1984 was greatly affected by the parallel and complementary diffusion of synthetic fibres (both used in shuttleless looms), just as diffusion of the looms was affected by the diffusion of open-end rotors[48].

Briefly, in a catching up approach it is recognised that the spread of new technology can be accomplished more freely by imitating countries than by innovating ones.

Adopters can buy the most up-to-date technologies without the burden of an existing stock of capital goods. International diffusion in imitating countries is thus likely to be faster than in innovating ones when:

1. Growth of demand is stronger in imitating countries; and therefore
2. Net investments are larger;
3. New technologies are incompatible with the existing capital stock;
4. The existing capital stock in imitating countries is inefficient, backward or even non-existent.

The distribution of advanced telecommunications networks aptly demonstrates the necessary connections between innovation diffusion, rates of economic growth and the specific features of the technology. Advanced telecommunications is also typical in its leapfrogging potential for less advanced countries and the ripple effect it produces within a country's economy: rapid modernisation of such strategic infrastructures as telecommunications has enormous advantages for the manufacturing and service industries, as we will see in Chapter 2. In fact a wave of technological and organisational innovations that enhance economic growth and set the stage for a variety of information and communication technologies is expected to enter countries equipped to receive it.

Towards an Integrated Approach

The debate on the causes and patterns of the international dissemination of innovations has been long and the results somewhat contrasting; that diffusion encourages growth and competitiveness nevertheless seems undeniable. To achieve strong, sustained growth, a country's ability to appropriate new processes rapidly, new capital goods and especially information appears to be at least as important as the ability to generate innovations.

Here we will summarise the results of our review and look at some of the principal reasons for the uneven patterns and timing of diffusion among countries.

a) *Investments*

Adoption of an innovated capital good or a process innovation depends first of all on the decision to invest[49]. Investment decisions in turn depend on factors such as: *i)* the availability of financial resources; *ii))* interest rates; *iii)* the age of existing capital goods; *iv)* expectations about future market trends; *v)* relative profitability of the new capital good compared to existing ones; *vi)* the degree of technical complementarity of innovated capital goods with existing ones; and *vii)* "animal spirits" – entrepreneurial optimism. Adoption can be delayed, even when large profits are possible, by a combination of a gloomy market outlook, high interest rates, credit rationing, young age composition (or *vintage*) of existing capital, and innovated technologies that must displace all existing capital stock to be successfully adopted.

Therefore a country's chances of quickly assimilating innovations are heavily influenced by the investment climate and the growth of demand.

b) *"Critical mass"*

According to the epidemic approach mentioned earlier, diffusion can be conceptualised as a shift of demand curves for innovated products and services resulting from a spread of information from early adopters to late ones[50]. Late adopters, who let others "test the waters" before adopting and are thus hampered by significant "search" costs, usually become aware of the higher profits of early adopting countries that had lower search costs. So the decision to adopt based on the perception of others' profitability (or loss) is very important. Early adopters can be considered risk-takers who, once they are numerous enough, create a "critical mass" – a gauge by which late adopters can judge adoption's benefits. Rapid formation of critical mass is in fact necessary for a spread of innovations generated abroad to domestic markets because of the opportunities it affords for weighing profitability against risk. This is distinctly true for adoption of radical innovations like advanced communications that require a complete overhaul of manpower skills and organisational routines. Adoption costs can be measured through evidence provided by a large number of users, and even the existence of this critical mass enhances profitability potential because of the widening pool of skilled manpower, enhanced technical assistance and the overall understanding of the new technology that comes from "learning by using"[51].

c) *The role of supply in innovation*

As many analysts have convincingly argued, information alone will not guarantee adoption in user countries. Supply, too, exerts pressure on potential users. Suppliers' attempts at persuasion through marketing and introduction of incremental innovations plays an important part in international diffusion[52]. First adopters are usually large firms with high wage rates but low search costs, and for them adoption is profitable even at elevated purchase prices. Late adopters, however, which are mostly smaller, with lower wages but higher search costs, can only adopt when the price of the innovated capital good compares favourably to alternative capital goods (such as the ones they already have). They can take advantage of a reduction in production costs brought about by economy of scale or learning by doing, or the use of incremental innovations.

The role of supply is therefore relevant in adoption decisions. Prices kept at high levels because of custom duties or tariff barriers combined with small domestic markets that lack the advantage of economies of scale in vertically related industries will delay the diffusion of innovated capital goods in adopting industries.

d)   *Technological features of the innovations*

Increasing empirical evidence shows that innovations based on "complex" technologies display diffusion patterns quite different from those within simpler, "self-contained" frameworks[53]. Innovations in elaborate systems like the new information technologies, with interconnections between telecommunications, computers, software, etc., and which demand much "learning by doing" and "learning by using", tend to be adopted only after the entire technological network, each innovated component in place, has been assembled and evaluated[54]. These cumulative technologies are viewed with a "wait and see" attitude and are rarely adopted before the formation of "critical mass". Once followers decide to go ahead, they can start from scratch and sidestep the problem of an older, depreciating technical system.

This "system" characteristic of certain new technologies – the interrelatedness of their innovated capital goods and intermediary inputs – may have an effect on the widening or narrowing of the technological gap between early and late adopting countries on international markets[55].

e)   *The behaviour of multinational firms*

The strategies of multinational firms appear to be evolving towards more interaction with local markets and more specialisation[56]. The shortened life-cycles of new products, heavier technological competition on international markets, the new information technologies and corresponding organisational innovations have all pushed multinational companies towards more global strategies and "network" organisation[57]. The model of the multinational company as a hive of affiliates that take innovations from home to exploit abroad (with significant time lags) is becoming more and more obsolete. Within the new worldwide corporations each affiliate operates with the most advanced technologies, albeit in a more narrowly defined product niche. Each affiliate is thus much more globally integrated and must attain world standards of efficiency and production.

This change could be very beneficial to developing countries that host these affiliates. As fast adopters and more specifically as lead users of advanced process technologies both in manufacturing and in the office, these firms can add incremental innovations that are appropriate to local conditions and help attain "critical mass" for adoption within the country.

In conclusion, the time it takes to import and then imitate a product can be therefore very short in countries that have:

- large firms, which can effectively process and analyse information they receive on new technologies, and that possess solid bargaining power;
- high exposure to international information flows through trade, or foreign direct investments by multinational companies;
- selective government protection and import duties, and subsequent low prices of key innovated products and services to local customers;
- strong economic growth, and particularly strong growth of demand and investments.

Small countries with fast growth and good investment potential, with export-oriented industrial strategies and a plan to integrate into international markets within specialised niches, are in a good position for high rates of diffusion.

NOTES AND REFERENCES

1. K.J. Arrow, "The Economic Implications of Learning by Doing", *Review of Economic Studies* (June 1962), pp. 155-173.

2. H.A. Simon, *Models of Bounded Rationality: Behavioral Economics and Business Organization*, MIT Press, Cambridge, 1982.

3. O.E. Williamson, *Markets and Hierarchies: Analysis and Anti-Trust Implications*, Free Press, New York, 1975.

4. J.E. Stiglitz, "Learning to Learn: Localized Learning and Technological Progress" in P. Dasgupta and P. Stoneman, eds., *Economic Policy and Technological Performance*, Cambridge University Press, London, 1987.

5. J.E. Stiglitz, "Economic Organization, Information and Development" in H. Chenery and T.S. Srinivasan, eds., *Handbook of Development Economics Vol. 1*, Elsevier, Amsterdam, 1988.

6. P. Dasgupta and J.E. Stiglitz, "Learning by Doing: Market Structure and Industrial and Trade Policies", *Oxford Economic Papers* (June 1988), pp. 246-268.

7. O.E. Williamson, *The Economic Institutions of Capitalism*, Free Press, New York, 1985.

8. P.M. Romer, "Increasing Returns and Long-Run Growth", *Journal of Political Economy* (October 1986), pp. 1002-1037.

9. M.F.G. Scott, *A New View of Economic Growth*, Clarendon Press, Oxford, 1989.

10. R.R. Nelson, "A Diffusion Model of International Productivity Differences in Manufacturing Industry", *American Economic Review* (December 1968), pp. 1219-1248.

11. R. Vernon, "International Investment and International Trade in the Product Cycle", *Quarterly Journal of Economics* (May 1966), pp. 198-207.

12. S. Hirsch, *Location of Industry and International Competitiveness*, Oxford University Press, London, 1967.

13. M.V. Posner, "International Trade and Technical Change", *Oxford Economic Papers* (October 1961), pp. 330-337.

14. R. Vernon, "International Investment and International Trade in the Product Cycle", *Quarterly Journal of Economics* (May 1966), pp. 198-207.

15. N. Kaldor, "Comment", *Review of Economics and Statistics 48*, (July 1962), pp. 118-120.

16. P.L. Swann, "The International Diffusion of Innovation", *Journal of Industrial Economics* (September 1973), pp. 61-69.

17. J. Tilton, *International Diffusion of Technology: The Case of Semiconductors*, The Brookings Institution, Washington, 1971.

18. T. Ozawa, "Government Control over Technology Acquisition and Firms' Entry into New Sectors: The Experience of Japan's Synthetic Fibre Industry", *Cambridge Journal of Economics* (June 1980), pp. 133-146.

19. R. Stobaugh, *Innovation and Competition: The Global Management of Petrochemical Products*, Harvard Business School University Press, Boston, 1988.

20. E. Mansfield, "Technical Changes and the Rate of Imitation", *Econometrica* (October 1961), pp. 741-766.

21. Z. Griliches, "Hybrid Corn: An Exploration in the Economics of Technological Change", *Econometrica* (October 1957), pp. 501-522.

22. R.R. Nelson and S.G. Winter, *An Evolutionary Theory of Economic Change,* Belknap Press of Harvard University Press, Cambridge, 1982.

23. H.A. Simon, *Models of Bounded Rationality: Behavioral Economics and Business Organization,* MIT Press, Cambridge, 1982.

24. C. Antonelli, "The International Diffusion of New Information Technologies", *Research Policy 3* (1986), pp. 139-147.

25. D. Allen, "New Telecommunications Services: Network Externalities and Critical Mass", *Telecommunications Policy* (September 1988), pp. 257-271.

26. D. David, "CLIO and the Economics of QWERTY", *American Economic Review* (May 1985), pp. 332-337.

27. P.A. David and J.A. Bunn, "The Evolution of Gateway Technologies and Network Evolution: Lessons from Electricity Supply History", *Information Economics and Policy 3* (1988), pp. 165-202.

28. E. Von Hippel, *The Sources of Innovation,* Oxford University Press, Oxford, 1988.

29. G.S. Maddala and P.T. Knight, "International Diffusion of Technical Change: A Case Study in the Oxygen Steel-Making Process", *Economic Journal* (September 1967), pp. 531-558.

30. E.K.Y. Chen, "Multinational Corporations and Technology Diffusion in Hong Kong Manufacturing", *Applied Economics* (June 1983), pp. 309-321.

31. R. Hsia, "Technological Change in the Industrial Growth of Hong Kong," in B.R. Williams, ed., *Science and Technology in Economic Growth,* Macmillan, London, 1973.

32. P. Stoneman, *The Economic Analysis of Technological Change,* Oxford University Press, London, 1983.

33. L. Nabseth and G.F. Ray, eds., *The Diffusion of New Industrial Processes,* Cambridge University Press, Cambridge, 1974.

34. P. David, "The Mechanization of Reaping in the Anti-Bellum Midwest" in N. Rosenberg, ed., *The Economics of Technological Change,* Penguin, Harmondsworth, 1971.

35. R.R. Nelson, M.J. Peck and E. Kalachek, *Technology, Economic Growth and Public Policy,* The Brookings Institution, Washington, D.C., 1967.

36. L. Lynn, "New Data on the Diffusion of the Basic Oxygen Furnace in the U.S. and Japan", *Journal of Industrial Economics* (December 1981), pp. 123-135.

37. C. Antonelli, "The International Diffusion of New Information Technologies", *Research Policy 3* (1986), pp. 139-147.

38. C. Antonelli, "Profitability and Imitation in the Diffusion of Process Innovations", *Rivista Internazionale di Scienze Economiche e Commerciali* (February 1990), pp. 109-126.

39. "The crucial importance of past investment in the size and age distribution of the established technology's capital stock and the existing commitment to the technology which is being displaced, in slowing the new technology, points also towards the phenomenon of intertechnology competition", L. Soete, "International Diffusion of Technology: Industrial Development and Technological Leapfrogging", *World Development* (March 1985), pp. 409-422.

40. M. Abramovitz, "Catching Up, Forging Ahead and Falling Behind", *Journal of Economic History* (June 1986), pp. 385-406.

41. Quoted from M. Abramovitz, "Following and Leading", in H. Hanusch, ed., *Evolutionary Economics, Applications of Schumpeter's Ideas,* Cambridge University Press, Cambridge (1988), p. 325.

42. G.F. Ray, "The Diffusion of New Technology: A Study in Ten Processes in Nine Industries," *National Institute Economic Review* (May 1969), pp. 40-83; *The Diffusion of Mature Technologies,* Cambridge University Press, Cambridge, 1984; and "Full Circle: The Diffusion of Technology", *Research Policy 1* (1989), pp. 1-18.

43. G.F. Ray, "Full Circle: The Diffusion of Technology", *Research Policy 1* (1989), pp. 1-18.

44. C. Antonelli, "The Role of Technological Expectations in a Mixed Model of International Diffusion of Process Innovations: The Case of Open-End Spinning Rotors", *Research Policy 5* (1989), pp. 273-288.

45. C. Antonelli, P. Petit and G. Tahar, "La Diffusion d'une nouvelle technique: L'application d'une innovation dans l'industrie textile", *Revue d'Economie Industrielle 48* (1989), pp. 1-15.

46. C. Antonelli, "The International Diffusion of Process Innovations and the Neotechnology Theory of International Trade", *Economic Notes 1* (1986), pp. 60-82.

47. M. Dalia, "Import Led Innovation: The Case of the Austrian Textile Industry", *Weltwirtschaftliches Archiv 3* (1988), pp. 550-565.

48. C. Antonelli, P. Petit and G. Tahar, "The Diffusion of Interdependent Innovations in the Textile Industry", *Structural Change and Economic Dynamics 2,* 1990.

49. C. Antonelli, P. Petit and G. Tahar, "Technological Diffusion and Investment Behavior: The Case of the Textile Industry", *Weltwirtschaftliches Archiv 4* (1989), pp. 782-803.

50. E. Mansfield, "Technical Changes and the Rate of Imitation", *Econometrica* (October 1961), pp. 741-766.

51. C. Antonelli, "Induced Adoption and Externalities in the Regional Diffusion of New Information Technology", *Regional Studies,* (February 1990), pp. 31-40.

52. P. Stoneman, *The Economic Analysis of Technological Change,* Oxford University Press, London, 1983, and C. Antonelli, "Profitability and Imitation in the Diffusion of Process Innovations", *Rivista Internazionale di Scienze Economiche e Commerciali,* (February 1990), pp. 109-126.

53. M. Frankel, "Obsolescence and Technical Change in a Maturing Economy", *American Economy Review* (September 1955), pp. 296-319.

54. C. Antonelli, "The International Diffusion of New Information Technologies", *Research Policy 3* (1986), pp. 139-147.

55. Together with organisational innovations based on new information technologies, one can think of new materials and biotechnology based on a variety of complementary innovations in the chemical, engineering and intermediary products industries.

56. R.E. Caves, *Multinational Enterprise and Economic Analysis,* Cambridge University Press, Cambridge, 1982.

57. C. Antonelli, "Multinational Firms, International Trade and International Telecommunications", *Information Economics and Policy 4* (1984): pp. 333-343; and C. Antonelli, "The Diffusion of an Organizational Innovation: International Data Communications and Multinational Industrial Firms", *International Journal of Industrial Organization 2* (1985), pp. 109-118.

Chapter 2

NEW INFORMATION TECHNOLOGIES AND ADVANCED TELECOMMUNICATIONS

Electronic Switching and Advanced Telecommunications

Advanced telecommunications is a technology that combines many different innovations, from digital and numeric switching and optical fibres to satellite transmission, all of which are undergoing rapid evolution. Nearly 90 per cent of all telecommunications equipment is produced in OECD countries, and can be divided into three broad categories: *public switching equipment,* which connects terminals and co-ordinates the entire telecommunications network; *terminals,* which are the end point of the network and include simple telephone sets, fax, telewriters, computers, modems, and private branch exchanges (PBX); and *transmission equipment,* which carries the signal between terminal stations and switching centres. Thirty per cent of total telecommunications equipment bought is switching equipment, and OECD countries produce more than 90 per cent of it. It is the technological and economic core of the telecommunications network[1]. Without an electronic switching infrastructure (both space division of the mid-1970s and fully digital switching introduced in 1974), complete access to trade and information, unrestricted by size of firms or location of users, is not possible[2].

The development of microelectronics and the introduction of peripheral units and terminals have multiplied the uses of computers, and the combination of computers and an advanced telecommunications infrastructure has allowed entire communications networks to spring up. The ability to link manufacturing plants within these networks and the possibilities of interfacing computers with other electronic systems and co-ordinating planning and production, have produced radical changes in a broad range of production processes. The electronic control of financial flows, inventory and distribution, and the creation of more flexible production systems, as well as on-line management of remote factories and offices, have become commonplace because of these innovations. Their timely adoption is today a prime element for high total productivity, lower costs and competitive advantage[3].

In brief, information technology can be described as technological and organisational innovations that exploit modern communications systems to co-ordinate activities. These innovations, which sharply reduce the production costs of a single operating sector, provide unprecedented control over the retrieval, use, storing and evaluation of information essential to the management and co-operation of diverse units in industry and elsewhere. Information technology has revolutionized the specialisation and division of labour among and within firms[4].

33

Because of information technologies' interdependence with an advanced telecommunications network, the decision to adopt these technologies and the speed of their diffusion is necessarily influenced by the physical characteristics of the network. When the new technology is incompatible – or "discontinuous" – and implementing the innovation requires the reorganisation of firms, production processes, plant lay-out and industrial structure of the work force, as well as the disposal of the existing stock of capital goods, then diffusion will most likely be slow. Anticipated replacement and substitution costs must be considered – they could shrink adoption profitability and delay diffusion. Conversely, then, continuity, or compatibility of new capital goods with existing ones, and compatibility of skills, organisational routines, processes and lay-out of plants and firms, will increase diffusion rates.

When a new technology requires an overhaul not only of existing technology but of the *modus operandi* that went with it, early adopters, who lack the experience and accumulated knowledge to operate the innovation, will face higher costs. Late adopters will benefit from the eventual availability of skilled manpower, technical assistance and maintenance, new operational procedures and manufacturing software.

It appears clear, then, that high levels of incompatibility between technologies will add to the delay of an innovation's adoption. Once more we see that only countries that can sustain strong levels of investment will be able to adopt the wide spectrum of interdependent innovations that make up a new technology.

Digital switching and optical fibre transmission are becoming more and more essential for the diffusion of the new information technologies, both from a technical and an economic point of view. Advanced telecommunications provide a vast array of specialised services that cannot be supplied by the older electromechanical system, and offer them to a much broader range of potential adopters, including small manufacturing and retailing firms. Also, the costs of telecommunications services within this system are much lower, as we shall see in a moment, and this enhances adoption profitability.

The use of microelectronics in the digitalisation of switching and transmission equipment is the main reason behind cost reductions: capital costs are smaller because 1) microelectronic components are cheaper than electromechanical ones; 2) there is less need for office and plant space; and 3) production processes become more flexible. Savings in maintenance costs can be dramatic. Automatic error detection and correction, two features of this new technology, greatly reduce manpower requirements.

In the United States the average purchase cost of electromechanical switching installed from 1970 to 1975 was about $600 per line, while at the beginning of the 1980s the average purchase cost of electronic switching was already $400 per line. Today the cost of a digital line has dropped to $200. When considered in real terms, discounted by inflation, the fall in prices is even more impressive.

Digital switching permits hefty savings in operating costs as well. Administrative tasks like accounting, billing and invoicing can be performed by specialised software programmes instead of personnel.

The Transition from Electromechanical to Electronic Switching

The transition from electromechanical switching to electronic is sharply affected by two specific technological features. First, the installation of electromechanical switching exchanges was accompanied by long-term capacity expectations; once the infrastructure was in place, lines could be added very cheaply and reach full working capacity within five to ten years. (The entire infrastructure could last well over fifty years.) Through the 1970s and part of the 1980s – until the fall of digital line prices in the mid-1980s – the average cost of adding capacity step by step to electromechanical switching was lower than the average total cost of installing electronic switching. Therefore earlier industrialised countries were likely to possess an elaborate electromechanical structure, especially if demand for telecommunications was strong in the early 1970s, and will face much higher costs than countries which have low absolute levels of telecommunications density or which invested little in the early 1970s.

Second, integrating the new electronic switching exchanges into an electromechanical infrastructure is much more expensive and technically complex than building a network of entirely electronic switching from scratch. The cost of integrating a fully digital exchange into an electromechanical network involves the purchasing and installation of gateway technologies that can add as much as 30 to 40 per cent of original digital lines' costs.

Countries that have low communications density and those with highly developed networks, then, have widely diverging cost outlooks and uneven incentives to adopt the new technology. The differentiated conditions of industrialised countries with a large installed electromechanical network, and developing countries with a small installed base and the general problems which arise in network transition, are analysed by Paul David as follows: "A superior network technology that arrived on the scene unanticipated could successfully challenge the incumbent if the latter's installed base was comparatively small in relation to the flow of gross additions to the facilities accessed by users of the network. The foregoing model simply abstracted from the question of the durability of additions made to the network, but evidently a high rate of physical depreciation of the installed base tends to undermine the position of a technologically inferior incumbent. Substantial durability of capital equipment embodying a technological standard – or even an overlapping positioning of successive generations of moderate durability – will reinforce it against efficient challengers."[5]

The Information Technologies

The convergence of telecommunications, information technologies and electronics into an advanced telecommunications system is having the following effects:
1. It is widening the scope of telecommunications networks and their possible applications;
2. It is creating new technical requirements for reliability, capacity, speed and compatibility that are difficult to add in a piecemeal fashion because of universality (interrelatedness) and sunk costs, two typical features of telecommunications networks;
3. The use of innovated telecommunications infrastructures are increasingly raising demand. Advanced communications services and related organisational innovations are so potentially profitable for firms that use them that the price elasticity of demand will decrease as diffusion progresses: even if purchase and adoption costs seem high, in the long run the benefits will be higher[6].

Moreover the physical quality of the telecommunications network – the quality of the signal, for example, which is better in electronic than in electromechanical lines – is becoming more and more important in the diffusion of the new information technologies. Telecommunications services are increasingly assuming the complementary role of a production factor technically inseparable from and essential to the product's performance[7].

The information technology appears to offer opportunities for "technology blending" to advanced telecommunications system users through capital-saving, organisational innovations. For example, users can take advantage of:

- more access to multi-sourcing to compare prices and services (as with the French Minitel system);
- global scope of procurement of products and services;
- enhanced customization of large-scale production;
- increased flexibility of production processes – more products can be made with the same equipment;
- reduction of stocks of intermediate products because of "just-in-time" methods of manufacturing, which allow production to take place upon the placement of the order;
- reduction of stocks of final product;
- reduced delivery time;
- less working capital;
- less invoicing delay;
- electronic franchising and retailing.

These organisational innovations are characterised by the following:

i) Once diffusion has taken place, the fixed costs of the equipment necessary to use the telecommunications network are very low. They can be as small as the price of a minicomputer that – as is usually the case – incorporates a modem.

ii) These low costs reduce adoption barriers especially to small firms, provided the supply of the telecommunications network is available to everyone (centrally produced and distributed) and a critical mass of users of new services has already been reached.

iii) Manufacturing and service industries can take advantage of organisational innovations while making little or no changes in existing equipment.

For these reasons, opportunities for technology blending offered by the information technology (using new service innovations and routines with older manufacturing processes) can be quite good:

"It makes an enormous difference whether a new technology requires the purchase and introduction of new equipment (especially when such equipment involves large fixed costs) or whether it can be added on or introduced as a modification to existing equipment. The prospects for technology blending will be very much shaped by the ease with which new technology can be introduced without having to scrap the old. In the extreme case, if a new technology requires the complete scrapping of an old one in order to take advantage of it, no blending is possible"[8].

Evidence has shown that in Italy, advanced telecommunications services and consequent organisational innovations have helped rejuvenate "low-tech" manufacturing and retailing sectors (in particular the garment, shoe and furniture industries), where traditional small-scale production was blended with sophisticated market segmentation and customization for large-scale selling strategies[9]. The overall increase in productivity of Italian firms that have adopted information technologies has been enormous, with salutary effects on the Italian economy[10].

Opportunities for technology blending appear to be strong particularly for developing countries, which often have small-scale production by small firms in traditional "low-tech" industries.

The building of an advanced telecommunications infrastructure, which can provide modern telecommunications services and induce organisational innovations, is therefore a powerful factor in the economic growth of all countries, both developing and industrialised ones. Some of the specific reasons for this are:

1. It will activate learning processes in the use and adaptation of advanced software, and this will create valuable "spillover" effects of skills accumulation and tacit knowledge.
2. The installation of a *universal* switching system will provide basic communications standards and protocols common to all potential users, and this will solve major compatibility problems that stem from "lock-in" effects. Lock-ins arise from the use of specialised networks based on leased lines and prevent or delay adoption of advanced communications systems by smaller firms.
3. The generalised supply of advanced telecommunications services – because of economies of scale in production – will drastically reduce production costs and consequently customers' prices. This will encourage adoption by small firms and less advanced users like rural communities, retailers, and cottage industries.
4. Extensive use of advanced telecommunications services helps form "critical mass", which radically improves the "user-value" of all communications services and thus the profitability of adoption. Not only do services become less costly, but more importantly, the complementary hardware (from fax machines to updated electronic systems used in manufacturing and businesses) become cheaper.

The diffusion of information and communication technologies through a modernised infrastructure can have powerful effects on the entire economic system. Large-scale use of these technologies directly influences productivity, cost effectiveness and competitiveness in industries with high levels of product differentiation and low levels of unit prices. In these industries, prompt availability of information about demand trends can boost competitive advantage. This is especially true in the textile, clothing, furniture, shoe and leather industries.

Growing evidence suggests that the international competitiveness of the textile, garment, toy and consumer electronics industries in Taiwan and the Philippines relies increasingly on advanced telecommunications services, which allow tight links to form between commercial distribution on American and European markets and local production.

These industries are often managed by large multinational corporations or organised within the framework of international sub-contracting relations based on long-term contracts, joint ventures or limited partnerships between small domestic management firms and multinational corporations.

Sectors with long and complex production processes involving the assemblage of varied components can reduce costs and enhance competitiveness by using information services. Knowledge about consumer needs can be retrofitted to diverse suppliers and permit inventory savings and more flexibility to meet shifting demand. This has been demonstrated by many engineering industries, particularly those that produce automobiles and household appliances.

This is the case for a few *chaebol* (large Korean groups) such as Daewo in motorcars and shipbuilding, and Lucky Gold Star and Samsung in electronics, who are quite advanced in using data networks to co-ordinate intra-group flows of intermediary goods and components, commercial and financial information and real-time reporting.

Generally all industries that produce goods for final consumers, and that experience considerable fluctuation in levels, composition and trends of a highly fragmented demand, can greatly benefit from the diffusion of information and communication services.

An advanced telecommunications system is also very important to service industries like banking, retailing, transportation, maintenance and insurance, where information and real-time communication are vital to the production process. A reduction in the costs of these services will indirectly enhance international competitiveness within the entire economic system, since lower market costs means lower costs for manufacturing firms exposed to international trade.

The extraordinary efficiency of the Hong Kong and Singapore financial markets is more and more based on the extensive use of such advanced telecommunications services. Moreover, the integration of these countries' local tourist industries into international tourism depends increasingly on the "capillary" availability of these telecommunications services for real-time airline and hotel reservations.

Last but not least, as the growing empirical evidence confirms, the availability of advanced telecommunications infrastructure and, consequently, the lower levels of tariffs for telecommunications services and their higher quality and reliability, are becoming important factors in affecting the decisions of location of new plants and offices by multinational global corporations. The quality, reliability and costs of telecommunications services are in fact becoming a major competitive factor for a growing number of manufacturing and service activities operated by multinational global corporations, located in a large variety of countries. The timely modernisation of the telecommunications infrastructure is thus likely to help sustain high rates of economic growth also by means of augmented inflows of foreign direct investments, and consequently, increase levels of output, employment, labour productivity and, most important, spillover of training and learning opportunities.

The case of the modernisation of the telecommunications infrastructure clearly shows how the economic analysis of radical technological change and its diffusion cannot be contained within the borders of a technology or an industry. The magnitude of "externality effects" and "spillovers" in downstream industries can be assessed only in a broader, inter-industrial perspective.

The fast adoption of advanced telecommunications is likely to initiate cumulative processes along the following lines:

1. The supply of high-quality, low-cost, universal telecommunications stimulates diffusion of modern information and communications services.
2. Extensive adoption of these services within and among firms favours the adoption of complementary processes and organisational innovations in plants and offices.
3. Increased diffusion of technological and organisational change helps improve productivity and reduce costs, and this sparks competitiveness and growth in a wide

range of downstream users, both industrial and tertiary, with significant cumulative effects.

4. Heightened competitiveness and growth encourage the demand for innovated products and services, and thus provides – because of increased division of labour, economies of scale and learning – further cuts in both production costs and market prices. This spurs the growth of industries that supply innovated goods.

5. Higher demand for innovated products and services will attract new competitors who push down price-cost margins and introduce incremental innovations, enlarging the "choice-set" and variety for potential users and adopters.

6. Finally, growing division of labour and specialisation will promote the creation of new advanced industries. These industries, in their search for new demand, will develop new "advanced" customers.

Such a cumulative process is rooted in the interaction of innovated products and services with downstream industries, intermediary markets and users in a dynamic environment where innovation diffusion is the driving force. This diffusion is accompanied by declining prices, both for intermediary products and final goods, "quality effects" and "quantity effects" (such as technological and organisational innovations), economies of scale, learning processes, dynamic externalities and, as mentioned above, increased division of labour.

This chain of dynamic effects is strengthened by the systemic character of new information and communication technologies. The heart of such a system of interdependent innovations lies in the meshing of technological and organisational innovations in manufacturing (such as flexible manufacturing systems and robotics) and information handling (such as computers, informatics and telematics) and advanced telecommunications services and equipment.

Consequently the adoption of new electronic switching by developing countries can generate tremendous "supply effects", driving the economic system towards intense adoption of an array of radical innovations.

It is essential to grasp the interdependency of information and communications technology diffusion to really understand the opportunities for growth in output, employment, productivity and competitiveness. This appears especially clear when analysis is focused on the "information and communications *filière*", i.e. the vertical relations of specialised industries that co-operate in the making of the finished product. In fact, one can argue that the diffusion of advanced telecommunications innovations within the *filière* behaved like a dynamic factor that created a network of interdependence among specialised activities. In this respect the diffusion of innovation has disclosed strong "dynamic externalities" as well as increasing returns that are the engine of growth in this area of the economic system.

The case of interdependent diffusion of technological innovations thus calls for a specific framework of analysis to comprehend the cause and effect of dynamic externalities that go far beyond the limits of given industries and entail such wide-ranging complementarities. This framework would bypass the static notion of externalities, and draw from the analysis of increasing returns and dynamic externalities by Allyn Young and Adam Smith[11]. In such a context a macroeconomic approach to the analysis of diffusion is indispensable.

The modernisation of a telecommunications infrastructure is an enormous task that can be accomplished only with intensive capital investment and technical skills. Replacing an electromechanical telecommunications infrastructure with a modern digital one demands extraordinary levels of investment for many reasons. First, because of digital technology's high levels of discontinuity with pre-existing technology it cannot be added on to the older infrastructure efficiently, therefore its adoption precludes the use of an older electromechanical network.

Second, the telecommunications infrastructure is characterised by complexity and interdependence among its components that prevent piecemeal substitution over time (and make it very expensive over space). Consequently full blocks of the infrastructure have to be replaced at one time.

Third, because the electromechanical infrastructure is so long-lived, wearing-out times are extraordinarily long – up to fifty years – and standard scrapping rates are low.

Fourth, the capital stock sunk in an electromechanical telecommunications infrastructure, measured at historic prices, is usually considered for many countries to equal GNP.

Fifth, investment levels in the electromechanical telecommunications infrastructure represent, in normal conditions, a significant chunk of gross fixed capital of a country, from 3 to 5 per cent, and from 0.5 to 1 per cent of GNP. The financial effort to modernise can quickly become oppressive, particularly in countries that have a fully established infrastructure.

Understanding the patterns and determinants of the diffusion of advanced telecommunications seems to require a firmer grasp of the relationship between investment decisions and adoption behaviour than has previously been indicated.

NOTES AND REFERENCES

1. Because of rapid advances in software, in the information processing and transmission technologies and fast increasing capacity of microprocessors, the borders between the private branch exchange technology and the central switching technology are becoming increasingly fuzzy especially for low density telecommunications networks. Consequently, the shares of central switching equipment on total telecommunications equipment are likely to decline over time. Figures in the text refer to large central switching equipment.

2. A.J. Roobeck, "Telecommunications: An Industry in Transition", in H.W. de Jong, ed., *The Structure of the European Industry*, Kluwer Academic Publishers, Dordrecht and Boston, 1988.

3. R.J. Sounders, J.J. Warford and B. Wellenius, *Telecommunications and Economic Development*, John Hopkins University Press, Baltimore, 1983; N.H. Leff, "Social Benefit Cost-Analysis and Telecommunications Investment in Developing Countries", *Information Economics and Policy 3* (1984), pp. 217-227; and H.H. Hudson and L.C. York, "Generating Foreign Exchange in Developing Countries: The Potential of Telecommunications Investments", *Telecommunications Policy* (September 1988), pp. 272-281.

4. C. Antonelli, ed., *New Information Technology and Industrial Change: The Italian Case*, Kluwer Academic, Dordrecht and Boston, 1988.

5. P.A. David, Some New Standards for the Economics of Standardization in the Information Age. In P. Dasgupta and P. Stoneman, eds., *Economic Policy and Technological Performance*, Cambridge University Press, Cambridge, 1987.

6. C. Antonelli, "The Diffusion of Information Technology and the Demand for Telecommunications Services", *Telecommunications Policy* (September 1989), pp. 255-264.

7. C. Antonelli, "Information Technology and the Derived Demand for Telecommunications Services in the Manufacturing Industry", *Information Economics and Policy 1* (1990), pp. 45-55.

8. N. Rosenberg, "New Technology and Old Debates", in A.S. Bhalla and D. James, eds., *New Technology and Development: Experiences in "Technology Blending"*, Lyenne Rienner Publishers, Boulder and London (1988), pp. 25-26.

9. A.S. Bhalla and D. James, eds., *New Technologies and Development: Experiences in Technology Blending*, Lyenne Rienner Publishers, Boulder and London (1988), p. 3.

10. C. Antonelli, ed., *New Information Technology and Industrial Change: The Italian Case*, Kluwer Academic, Dordrecht and Boston, 1988.

11. See the celebrated quote:

"Adam Smith's famous theorem amounts to saying that the division of labor depends in large part on the division of labor. This is more than mere tautology. It means, if I read its significance rightly, that the counter forces which are continually defeating the forces which make for economic equilibrium are more pervasive and deeply rooted in the constitution of the modern economic system than we commonly realise. Not only new or adventitious elements coming in from the outside, but elements which are permanent characteristics of the ways in which goods are produced make continuously for change. Every important advance in the organization of production regardless of whether it is based upon anything which in a narrow or technical sense,

41

would be called a new 'invention" or involves a fresh application of the fruits of scientific progress to industry, alters the conditions of industrial activity and initiates responses elsewhere in the industrial structure which in turn have a further unsettling effect. Thus change becomes progressive and propagates itself in a cumulative way". A. YOUNG (1928) p. 533.

Chapter 3

THE EMPIRICAL EVIDENCE ON THE INTERNATIONAL DIFFUSION OF ADVANCED TELECOMMUNICATIONS

The Levels of Diffusion

Diffusion levels of electronic switching[1] – the key element of advanced telecommunications – were quite high by the end of the 1980s. By 1987 only a small number of countries had not yet been able to adopt the new electronic technology and modernise a significant part of their telecommunications infrastructure. Data gathered in Table 1 shows that out of 50 countries[2] (for which data were available) with more than 200 000 lines, only ten had not begun diffusion of electronic switching by 1987 – in those countries, penetration levels were equal to or below 10 per cent of total switching stock.

When compared to the international dissemination of other major technologies like continuous casting, basic oxygen steel, float glass, numerically controlled machine tools, shuttleless looms, and open-end spinning, such a level of international diffusion only fifteen years after introduction seems extremely high.

Even more unusual is the composition of the late adopters. Here we find "unexpected" highly industrialised countries like Belgium and the Federal Republic of Germany with penetration of electronic switching capacity below 1.5 per cent of total exchange lines, while Italy has penetration levels of 11.9 per cent. Some late adopters with low penetration levels are "expected" late industrialising countries like Algeria (1.4 per cent), Argentina (6.5 per cent), Greece (0.0 per cent), Venezuela (8.3 per cent), Yugoslavia (0.0 per cent) and a few others.

In the other forty countries diffusion was much stronger. In only five countries, however, was diffusion close to completion (i.e. penetration levels over 70 per cent): Norway (100 per cent), Saudi Arabia (93.4 per cent), the United States (76.2 per cent), South Korea (70.3 per cent) and France (69.4 per cent). It is interesting to note that among the five top adopters, two are newly industrialising countries.

Following Maddala and Knight's methods[3] we will classify countries into the following categories:

1. Leaders:
   countries whose ratio of electronic lines to total switching capacity was greater than 50 per cent in 1987;
2. Fast Followers:
   countries whose ratio was between 35 per cent and 50 per cent in 1987;
3. Slow Followers:
   countries whose ratio was between 20 per cent and 35 per cent;

4. Laggards and Non-Adopters:
  countries whose ratio was below 20 per cent, and countries with no electronic switching in 1987.

Thirteen countries enter the class of leaders, including only four OECD countries (USA, Canada, France and Norway). All the other leaders are Asian countries except for Morocco.

Ten countries are classified as fast followers, of which four in the OECD region (United Kingdom, the Netherlands, Ireland and Japan) and three are from Latin America. The rest are Asian countries, both in the Near and Far East.

Seven countries are in the category of slow followers, three of which are OECD members. Seventeen countries belong to the class of laggards. OECD countries are in the majority with seven entries: Austria, Belgium, Denmark, Federal Republic of Germany, Finland, Italy and Spain. Of five non-adopters, three are OECD members: Australia, Greece and Yugoslavia.

The results of these classifications are remarkable: most of Europe and Latin America fall markedly behind North America and Asia, especially Far East Asia.

The comparison of these results with those of penetration levels of the basic oxygen furnace studied by Maddala-Knight[3] yields even more surprising figures. In this study only one non-OECD member, Brazil, entered the leaders' category, in the company of nine OECD countries. Only two countries, both in the OECD, were fast followers. Out of six slow followers, four were OECD Members. One OECD country accompanied nine non-OECD members (not including socialist countries) in the laggards and non-adopter categories.

Apparently the international diffusion of advanced telecommunications in recent years has not only been very fast, it has been faster in late industrialising countries in Far East Asia than in OECD countries.

In 1977 the diffusion of electronic switching had reached only thirteen countries out of forty-eight (see Table 3) and in fact a significant share of total exchange lines had been modernised only in Canada, Hong Kong, Malaysia, the United Kingdom and the United States. These delays in the date of origin of the diffusion process across countries is more in keeping with traditional expectations than the range of penetration levels in 1987. Evidently most developing and newly industrialising countries, both in Latin America and Far East Asia, started later than traditional highly industrialised leaders and yet experienced much faster growth.

The most striking case is that of South Korea, where diffusion reached the 10 per cent benchmark in 1981, four years after the United States, but then reached 70.3 per cent in 1987. In Malaysia, electronic switching accounted in 1977 for 7.4 per cent of total capacity of switching lines – as much as in the United Kingdom but less than in Canada – but by 1987 it had reached 64.3 per cent, a higher level than in both countries. Singapore started from a low of 5 per cent in 1977, to arrive at 64.5 per cent in 1987.

In Latin America, a similar spectacular rise in diffusion occurred in Chile, from only 1.4 per cent in 1982, the year of introduction, to 45.8 per cent in 1987. Peru started late in 1980 but had risen to 41.1 per cent diffusion by 1987.

The classification of countries according to 1977 penetration levels into the three categories – leaders (electronic switching equal to or greater than 10 per cent of total switching capacity), adopters (electronic switching higher than zero), and laggards (including non-adopters) – produces even more startling results. Only two OECD countries had modernised at least 10 per cent of switching capacity: the United States and Canada. Four OECD counties (Denmark, France, Sweden and the United Kingdom) had adopted their first electronic switching, along with seven late industrialising countries: Mexico in Latin America, Hong Kong, Malaysia, the Philippines and Singapore in Far East Asia, and Malta and South Africa. Apparently many late industrialising countries adopted at the very beginning of the diffusion process.

Combining the results of the classifications of Tables 2 and 3 suggests that the economic features of early adopters not belonging to the OECD were the following:

- small size of the economy (as reflected in the number of island states among the early adopters: Hong Kong, Malta, Singapore and Taiwan);
- low levels of pre-existing telecommunications infrastructure at the end of the 1970s;
- a lack of vertical integration – both formal and informal – with hardware manufacturers.

The data in Tables 1 to 4 suggest that penetration levels of advanced telecommunications are irregularly distributed throughout countries. Not only have many newly industrialising countries like Hong Kong, South Korea, Malaysia, Singapore and Thailand adopted advanced telecommunications earlier and more radically than most of Europe, but in many developing countries like Chile, Colombia, Morocco, Peru and Turkey where diffusion began later than in Canada, the United Kingdom or United States, the process spread at even faster rates. In a few years these developing countries' diffusion had reached the penetration levels of countries whose diffusion process had begun five to ten years earlier. European countries and Japan therefore can be characterised as both late starters and slow diffusers.

The Rates of Diffusion

An assessment of the rates of diffusion is as important as diffusion levels in understanding international differences in the absorption of the new telecommunications technology. The traditional methodology for studying diffusion rates of innovations is based on logistic models drawn from the epidemic approach (see Annex 1). The logistic curve can be specified as follows:

[i] $\qquad \log y_{it} = a + b_i t$

where $y_{it} = \dfrac{P_{it}}{K - P_{it}}$

P_{it} = the share of electronic lines in total lines in country i at year t
K = the ceiling level of potential adopters set equal to 100.

Equation i tests for data of Table 1 in the period 1977-1986. Results of GLS estimates are listed in Table 5.

The economic strength of the epidemic model is confirmed. Total variance explained is very high. The test of the epidemic model is a reliable gauge of the rates of diffusion.

As we have seen, diffusion rates from 1977 to 1986 were very high for most countries. Newly industrialising countries, particularly, performed extremely well. Results of Table 5 allow us to classify countries according to speed of diffusion of advanced telecommunications, as measured by the estimated value of the parameter b_i of equation i, into three categories[4]:

Forging ahead:

countries with fast diffusion rates (i.e. where the estimated value of the parameter b is larger than 1);

Fast adopters:

countries with average diffusion rates (where the estimated value of the parameter b falls between 0.5 and 1);

Falling behind:

countries with slow diffusion rates (where the estimated value of the parameter b is below 0.5).

The results of these classifications again are surprising. Only three OECD countries have been "forging ahead", with two countries from Latin America, two from the Far East, two from the Near East, and Morocco from Africa. Seven Latin American countries and six OECD countries are in the class of fast adopters. OECD countries constitute the absolute majority of countries falling behind.

The causes of these phenomena have not yet been firmly established. It does seem, however, that the case of international diffusion of advanced telecommunications supports the hypothesis of catching up and leapfrogging rather than the traditional concept of diffusion among imitating countries that mimics that of innovating countries. It is especially inappropriate to use the product-cycle approach here when one considers that the bulk of world innovative capacity of the telecommunications industry is concentrated in the advanced countries.

International Diffusion of Technological Capacity in Advanced Telecommunications

The evidence analysed so far shows that diffusion of electronic switching was very fast in the 1980s, faster in many late industrialising imitating countries than in most OECD innovating ones. Data in Table 7 indicate the distribution of total patents issued in the telecommunications industry between 1971 and 1988 by the US Patent Office. Out of 19 162 patents granted, US firms (and individuals) accounted for 65.6 per cent, corresponding to a density of 510 patents per million inhabitants. US leadership in telecommunications innovations equals its leadership in diffusion of the technology within the country. France is in a similar category with 1 013 patents, or 3 per cent of total patents issued, and 18.2 patents per million inhabitants. Canada has 572 patents, which amount to 4 per cent of the total, and 22.3 patents per million people, while the United Kingdom has 668 patents, 3.5 per cent of the total, and 12.2 patents per million inhabitants.

This correlation between innovation capacity and diffusion capacity is not observed, however, for many countries like Japan, whose nationals account for 2 017 patents, comprising 10.5 per cent of total patents and 16.64 patents per million inhabitants; Germany (1 104 patents, 5.7 per cent, 18.0 per capita), Sweden (188 patents, 1 per cent, 22.4 per capita), and Switzerland (169 patents, 0.9 per cent, 25.5 per capita). Less remarkable results are found in Austria, Belgium and Italy where both penetration levels in 1977 and 1987 and diffusion levels are well below the average.

It is interesting to note that many early adopting countries and many fast adopting countries, including Chile, Colombia, Hong Kong, Korea, Malaysia, Morocco, Peru, the Philippines, Singapore, Thailand and Turkey, have never received one patent from the US Patent Office.

A correlation between innovation capacity and internal diffusion capacity appears to be missing in the case of advanced telecommunications. However, a slight correlation was found at the beginning of the diffusion process between innovation capacity measured by the number of telecommunications patents granted in the period 1971-1976, and internal diffusion capacity, measured by 1977 penetration levels. The share of total patents granted to the United States, an early adopter, was slightly larger in 1976 than that of the period 1977-1988 (see Tables 8 and 9). In that year, US nationals accounted for 71.2 per cent of total patents, as opposed to only 62.2 per cent in the entire period from 1977 to 1988. During this same period, Japan, a late adopter, doubled its share of patents from 6.1 per cent to 12.8 per cent. Japan's case was typical of smaller late adopters like Austria, Denmark, Finland, Ireland and New Zealand, which received from 70 to 80 per cent of their patents in the period 1977-1987, when their electronic diffusion began.

NOTES AND REFERENCES

1. CMEA countries are not considered here.
2. Here we refer to electronic technology as both space-division switching, introduced in the late 1960s, and digital switching, introduced in 1974. See AT&T (1975).
3. G.S. Maddala and P.T. Knight. "International Diffusion of Technical Change: A Case Study in the Oxygen Steel-Making Process", *Economic Journal* (September 1967), pp. 531-558.
4. M. Abramovitz, "Following and Leading", in H. Hanusch, ed., *Evolutionary Economics,* Cambridge University Press, Cambridge, 1988.

Table 1. Total exchange lines and share of electronic switching capacity in 1987

	Total lines (000'S)	Electronic lines %		Total lines (000'S)	Electronic lines %
Canada	11 399	55.8	Spain	10 236	15.9
USA	118 400	76.2	Sweden	5 373	30.0
Costa Rica	240	16.8	Switzerland	3 499	31.3
Cuba	307	0.0	United Kingdom	23 183	48.4
Mexico	3 987	15.6	Yugoslavia	3 028	0.0
Uruguay	312	11.5	Malta	110	58.5
Argentina	3 034	6.5	Australia	6 816	0.0
Chile	608	45.8	Indonesia	765	0.0
Colombia	1 967	35.8	Malaysia	1 131	64.3
Ecuador	437	25.6	New Zealand	1 376	29.1
Peru	462	41.1	Philippines	444	42.2
Venezuela	1 678	8.3	Singapore	875	64.5
Algeria	634	1.4	Thailand	901	50.7
Morocco	266	50.0	Taiwan	4 909	26.6
S. Africa	2 495	15.2	Hong Kong	1 988	63.5
Tunisia	227	44.4	Japan	49 976	36.1
Austria	2 906	1.3	S. Korea	8 695	70.3
Belgium	3 367	1.0	Sri Lanka	97	57.4
Denmark	2 827	10.8	Iran	1 481	5.7
Finland	2 365*	15.5*	Iraq	787	58.8
France	24 803	69.7	Israel	1 392	28.5
Germany	27 512	1.5	Jordan	193	97.5
Greece	3 465	0.0	Pakistan	606	9.2
Ireland	798	45.9	Saudi Arabia	1 504	93.4
Italy	19 104	11.9	Syria	471	24.7
Netherlands	6 234	49.3	Turkey	3 702	36.2
Norway	1 984	100.0	UAE	261	42.2

* Data for 1986.

Source: AT&T, The World Telephones, 1989, and Italtel.

Table 2. Extent of adoption of electronic switching in 1987

Latin America	OECD	Far East	Near East & Africa
LEADERS 50%			
	Canada USA France Norway	Singapore Thailand Malaysia Hong Kong S. Korea Sri Lanka	Saudi Arabia Morocco Iraq Jordan Malta
FAST FOLLOWERS 35%-50%			
Chile Colombia Peru	United Kingdom Ireland Netherlands Japan Turkey	Philippines	UAE
SLOW FOLLOWERS 20%-35%			
Ecuador	Sweden New Zealand Switzerland	Taiwan	Israel Syria
LAGGARDS 20%			
Costa Rica Mexico Uruguay Venezuela Argentina	Belgium Austria Denmark Germany Italy Spain Finland		S. Africa Algeria Pakistan Iran
NON ADOPTERS			
	Australia Yugoslavia Greece	Indonesia	

Table 3. Extent of adoption of electronic switching in 1977

Latin America	OECD	Far East	Near East & Africa
LEADERS 10%			
	USA Canada		
FOLLOWERS			
Mexico	Denmark France Sweden United Kingdom	Singapore Malaysia Hong Kong Philippines	S. Africa Malta
LAGGARDS			
Costa Rica El Salvador Panama Argentina Chile Colombia Ecuador Peru Venezuela	Austria Belgium Finland Germany Greece Netherlands Norway Switzerland Yugoslavia Australia New Zealand Japan Turkey	Thailand Taiwan S. Korea	Algeria Morocco Tunisia Sri Lanka Iran Iraq Israel Jordan Pakistan Saudi Arabia Syria UAE

Table 4. Evolution of the share of electronic switching capacity on total exchange lines

	1977	1978	1979	1980	1981	1982	1983	1984	1985	1986	1987
Canada	14.1	16.1	20.1	28.8	26.3	25.6	31.2	42.9	47.7	51.4	55.8
USA	10.2	19.6	29.0	35.3	42.3	48.0	51.3	59.3	66.8	71.8	76.2
Costa Rica	0.0	0.0	0.0	0.0	0.0	0.0	0.0	0.0	8.7	15.9	16.8
El Salvador	0.0	0.0	0.0	0.0	0.0	3.0	7.0	18.4	25.1	24.3	28.3
Mexico	1.7	2.1	2.3	2.4	2.5	2.6	2.6	2.9	3.0	13.0	15.6
Panama	0.0	0.0	4.1	5.4	9.6	11.0	20.0	19.2	20.9	24.5	26.7
Argentina	0.0	0.0	0.0	1.1	2.2	3.1	3.5	3.5	3.9	4.1	6.5
Chile	0.0	0.0	0.0	0.0	0.0	1.4	10.0	32.1	30.8	39.0	45.8
Colombia	0.0	0.0	3.1	2.9	7.8	8.5	9.4	21.2	20.9	36.8	35.8
Ecuador	0.0	0.0	0.0	0.0	0.8	1.6	2.2	8.0	14.0	22.4	25.6
Peru	0.0	0.0	0.0	9.4	10.0	15.6	25.1	30.0	35.0	39.3	41.1
Venezuela	0.0	0.0	0.0	0.0	0.4	1.0	3.1	3.6	3.9	4.0	8.3
Algeria	0.0	0.0	0.0	0.0	0.0	0.0	0.0	1.0	1.0	1.0	1.4
Morocco	0.0	0.0	0.0	0.0	0.0	0.0	18.6	23.3	28.9	49.1	50.0
S. Africa	0.8	0.9	1.4	1.7	3.4	2.5	4.8	5.0	8.7	8.8	15.2
Austria	0.0	0.0	0.0	0.0	0.0	0.0	0.0	0.0	0.0	0.6	1.3
Belgium	0.0	11.6	14.5	17.2	21.3	25.6	33.3	38.2	41.1	45.5	47.9
Denmark	0.3	3.6	3.5	3.5	3.4	3.4	3.6	4.4	6.1	6.8	10.8
Finland	0.0	0.0	0.0	0.0	0.0	0.1	1.6	4.0	11.0	15.5	20.0
France	2.0	3.0	6.4	12.8	18.3	26.5	34.0	49.7	56.2	63.2	69.7
Germany	0.0	0.0	0.3	0.5	0.7	1.2	1.2	1.2	1.4	1.5	1.5
Greece	0.0	0.0	0.0	0.0	0.0	0.0	0.0	0.0	0.0	0.0	0.0
Ireland	0.0	0.0	0.0	0.0	0.0	0.0	5.3	16.1	25.9	37.4	45.9
Italy	0.0	0.0	0.5	0.0	1.8	3.4	1.0	2.3	4.6	8.2	11.9
Spain	0.2	0.2	0.2	0.5	1.8	3.6	6.8	8.3	9.2	13.0	15.9
Sweden	0.0	0.2	0.0	1.8	6.5	9.1	10.3	14.1	18.0	23.0	30.0
Switzerland	0.0	0.0	0.0	0.5	0.0	0.0	0.0	5.0	10.0	20.7	31.3
United Kingdom	7.0	10.0	25.5	25.5	14.9	23.8	30.0	35.5	40.0	42.6	48.4
Yugoslavia	0.0	0.0	0.0	0.0	0.0	0.0	0.0	0.0	0.0	0.0	0.0
Malta	25.5	29.3	32.9	35.1	37.4	38.7	52.6	53.0	54.0	57.2	54.2
Australia	0.0	0.0	0.0	0.0	0.0	0.0	0.0	0.0	0.0	0.0	0.0
Indonesia	0.0	0.0	0.0	0.0	0.0	0.0	0.0	0.0	0.0	0.0	0.0
Malaysia	7.4	14.9	21.0	27.1	26.1	30.1	33.1	39.1	48.6	56.7	64.3
New Zealand	0.1	0.5	1.0	2.2	0.0	0.4	2.6	9.3	8.9	15.3	29.1
Philippines	4.0	5.0	11.7	26.1	8.0	11.4	12.1	35.1	39.7	40.7	42.2
Singapore	0.0	0.0	0.0	0.0	36.5	44.6	42.3	50.2	52.4	57.2	64.5
Thailand	0.0	0.0	0.0	0.0	0.0	0.0	2.6	9.0	27.1	45.5	50.7
Taiwan	0.0	0.0	0.0	0.0	0.0	0.0	14.7	17.2	20.9	25.2	26.6
Hong Kong	14.3	17.1	22.0	25.9	32.9	38.2	42.1	42.2	43.1	54.7	63.5

52

Sri Lanka	0.0	14.0	26.1	28.0	30.0	49.5	50.0	51.0	52.7	57.4
Israel	0.0	0.0	0.0	0.2	2.0	13.1	16.0	21.0	24.8	28.5
Pakistan	0.0	1.9	1.6	1.5	1.1	1.3	1.2	1.2	10.2	9.2
Saudi Arabia	0.0	0.0	0.0	0.0	0.0	0.0	82.8	97.8	93.3	93.4
Syria	0.0	0.0	0.0	0.0	0.0	0.0	0.0	0.0	25.0	24.7
Turkey	0.0	0.0	0.0	0.0	0.2	3.2	8.8	11.4	17.7	36.2
UAE	0.0	0.0	0.0	0.0	0.0	0.0	35.8	36.3	47.5	42.2

Source: AT&T and Italtel.

Table 5. **Results of the GLS estimates of equation** i

	a	b	$\bar{R}^2$
Canada	−3.238 (17.287)	0.203 (13.496)	0.947
USA	−3.825 (15.714)	0.301 (15.381)	0.963
Costa Rica	−16.897 (6.308)	0.807 (3.740)	0.608
El Salvador	−18.263 (9.371)	1.091 (6.953)	0.843
Mexico	−5.624 (9.563)	0.189 (4.001)	0.601
Panama	−12.188 (5.497)	0.750 (4.198)	0.624
Argentina	−13.490 (7.645)	0.707 (4.974)	0.703
Chile	−19.086 (9.774)	1.176 (4.475)	0.845
Colombia	−12.976 (6.750)	0.806 (5.208)	0.723
Ecuador	−17.031 (13.234)	0.993 (9.584)	0.900
Peru	−15.274 (6.419)	0.991 (5.171)	0.720
Venezuela	−15.508 (11.799)	0.819 (7.734)	0.854
Morocco	−19.691 (7.770)	1.194 (5.852)	0.768
S. Africa	−7.008 (31.457)	0.302 (16.840)	0.965
Belgium	−8.258 (3.618)	0.546 (2.969)	0.438
Denmark	−5.852 (7.887)	0.213 (3.563)	0.539
Finland	−17.696 (12.218)	0.979 (8.388)	0.874
France	−7.027 (23.941)	0.483 (20.431)	0.976
Germany	−11.178 (8.262)	0.474 (4.532)	0.642
Ireland	−19.131 (8.519)	1.144 (6.261)	0.792
Italy	−17.073 (10.164)	0.884 (6.534)	0.806
Spain	−12.960 (8.658)	0.732 (6.073)	0.782
Sweden	−17.313 (8.311)	0.871 (4.339)	0.761

Table 5 *(cont.)*

	a	b	$\bar{R}^2$
Switzerland	−18.212 (7.549)	0.974 (5.011)	0.706
United Kingdom	−3.719 (9.044)	0.218 (6.599)	0.809
Malta	−2.011 (11.948)	0.141 (10.451)	0.915
Malaysia	−3.349 (15.256)	0.262 (12.577)	0.940
New Zealand	−17.857 (11.826)	1.015 (8.345)	0.872
Philippines	−10.469 (13.830)	0.654 (10.727)	0.919
Singapore	−5.261 (9.200)	0.368 (7.990)	0.826
Thailand	−19.451 (9.000)	1.148 (6.596)	0.809
Hong Kong	−3.217 (19.480)	0.214 (16.133)	0.962
Japan	−18.935 (6.803)	1.060 (4.728)	0.681
S. Korea	−17.299 (9.573)	1.164 (7.996)	0.862
Sri Lanka	−16.301 (8.621)	0.980 (7.110)	0.891
Israel	−17.652 (11.446)	1.063 (8.554)	0.878
Pakistan	−11.411 (6.332)	0.551 (3.794)	0.572
Turkey	−18.189 (11.790)	1.044 (8.404)	0.874

Table 6. **An international classification according to the rates of diffusion of electronic switching**

Latin America	OECD	Near East & Africa	Far East
FORGING AHEAD			
El Salvador Chile	New Zealand Ireland Japan Turkey	Israel Morocco	Thailand S. Korea Sri Lanka
FAST ADOPTERS			
Costa Rica Panama Argentina Colombia Ecuador Peru Venezuela	Belgium Finland Italy Spain Switzerland Sweden	Pakistan	Philippines
FALLING BEHIND			
Mexico	Canada USA Denmark France Germany United Kingdom	S. Africa Malta	Malaysia Singapore Hong Kong

The classification is made according to the results of table 5.

Table 7. **Total patents issued in the period 1971-1988 in telecommunications**

	Total patents	%	1987 population in thousands	Density per 10 million
USA	12 566	65.57	246 113	510.0
Japan	2 017	10.52	122 458	16.4
Germany	1 104	5.76	61 170	18.0
France	1 013	5.28	55 514	18.2
UK	668	3.48	54 731	12.2
Canada	572	2.98	25 625	22.3
Netherlands	301	1.57	14 713	20.4
Italy	277	1.44	57 406	5.0
Sweden	188	0.98	8 382	22.4
Switzerland	169	0.88	6 619	25.5
Belgium	67	0.34	9 924	6.7
Austria	39	0.20	7 573	5.1
Norway	30	0.15	4 198	7.1
Ireland	26	0.13	3 543	7.3
Denmark	12	0.06	5 129	2.3
Soviet Union	12	0.06	278 000*	0.43
Finland	8	0.04	4 941	1.6
New Zealand	6	0.03	3 307	2.0
Poland	6	0.03	37 572*	1.6
Hungary	6	0.03	10 604*	5.6
Czechoslovakia	6	0.03	15 587*	3.8
Greece	5	0.02	9 980	0.5
Spain	5	0.02	39 092	0.12
Brazil	2	0.01	141 452	0.01
Romania	2	0.01	22 700*	0.88
Luxembourg	1	–	–	–
TOTAL	19 108	100		

Source: Von Tunzelman's personal data file (1989).

Table 8. **Total patents issued in the period 1971-1976 in telecommunications**

	Total patents	%
USA	4 702	71.2
Japan	402	6.1
Germany	382	5.8
France	282	4.3
United Kingdom	222	3.3
Canada	181	2.7
Netherlands	88	1.3
Italy	98	1.5
Sweden	78	1.2
Switzerland	61	0.9
Belgium	34	0.5
Austria	9	0.1
Norway	16	0.2
Ireland	5	0.0
Denmark	2	0.0
Soviet Union	6	0.0
Finland	2	0.0
New Zealand	2	0.0
Poland	2	0.0
Hungary	2	0.0
Czechoslovakia	4	0.0
Greece	4	0.0
Spain	3	0.0
Brazil	0	0.0
Romania	2	0.0
Luxembourg	0	0.0
TOTAL	6 590	100

Source: Von Tunzelman (1989).

Table 9. **Total patents issued in the period 1977-1978 in telecommunications**

	Total patents	%
USA	7 864	62.6
Japan	1 615	12.8
Germany	722	5.7
France	731	5.8
United Kingdom	446	3.5
Canada	391	3.1
Netherlands	213	1.7
Italy	179	1.4
Sweden	110	0.8
Switzerland	108	0.8
Belgium	33	0.3
Austria	30	0.3
Norway	14	0.1
Ireland	21	0.1
Denmark	10	0.1
Soviet Union	6	0.1
Finland	6	0.1
New Zealand	4	0.1
Poland	4	0.1
Hungary	4	0.1
Czechoslovakia	2	0.1
Greece	1	0.1
Spain	2	0.1
Brazil	2	0.1
Romania	0	0.1
Luxembourg	1	0.1
TOTAL	12 518	

Source: Von Tunzelman (1989).

Table 10. **Patents delivered in telecommunications**

Including modulators, demodulators, detectors

	1971	1972	1973	1974	1975	1976	1977	1978	1979	1980	1981	1982	1983	1984	1985	1986	1987	1988
USA	676	738	883	799	845	761	798	731	717	605	450	635	574	568	546	680	788	772
Japan	44	45	83	72	72	86	111	110	129	131	88	148	120	137	120	125	181	215
Germany	71	78	61	55	60	57	67	64	65	43	46	66	50	69	52	59	71	70
United Kingdom	49	37	51	37	25	23	46	39	41	32	13	35	39	34	32	31	46	58
Switzerland	4	7	13	15	10	12	6	6	11	21	7	7	2	10	4	14	11	9
France	30	33	57	51	63	48	57	51	55	57	42	43	57	59	66	65	96	83
Netherlands	12	14	12	16	16	18	10	16	15	21	8	16	20	22	17	22	29	17
Sweden	18	16	7	12	11	14	10	18	11	15	6	5	8	8	3	3	11	12
Italy	12	14	22	19	17	14	8	18	21	13	10	10	18	23	14	11	21	12
Denmark	0	0	0	1	1	1	0	1	0	1	0	2	0	3	1	1	2	2
Belgium	5	10	10	6	2	0	7	4	3	3	1	1	5	2	2	0	3	3
Canada	16	32	30	37	31	35	30	29	31	38	24	25	27	39	23	37	49	39
Austria	2	2	0	3	1	0	2	6	2	3	1	0	4	2	2	1	3	4
Finland	0	0	1	1	0	2	0	0	0	0	0	3	1	0	0	0	0	2
Greece	0	1	0	0	1	1	0	0	0	0	0	0	0	1	0	0	0	0
Ireland	0	0	2	0	1	0	3	2	0	0	0	0	0	2	2	0	5	5
Norway	2	5	3	3	2	2	2	2	1	0	0	1	1	2	0	1	2	2
New Zealand	0	0	0	0	0	0	0	0	0	0	0	0	0	0	0	0	0	0
El Salvador	0	0	0	1	2	1	0	0	1	0	0	0	1	1	0	0	0	0
Soviet Union	0	1	3	1	1	0	0	1	0	0	0	0	0	0	0	0	2	0
Poland	0	0	2	0	0	0	0	0	0	0	0	0	2	0	0	0	0	0
Hungary	1	0	0	1	0	0	0	0	2	0	1	0	0	0	0	0	0	3
Romania	0	0	0	0	0	0	0	0	0	0	0	0	0	0	0	0	0	0
Luxembourg	0	0	0	0	0	1	0	0	0	0	0	0	1	0	0	0	0	0
Czechoslovakia	3	0	0	0	0	1	0	1	1	0	0	0	0	0	0	0	0	0

Source: Von Tunzelman (1989).

Chapter 4

THE ROLE OF SUPPLY, MARKET STRUCTURE
AND INDUSTRIAL ORGANISATION

Technological advance in telecommunications hardware has been impressive, giving rise to sweeping changes in production and costs as well as in the structure of the industry[1]. The appearance of microelectronics and their use in electronic switching dramatically shook the technological foundations of an industry that had been relying on electromechanics since the beginning of the century.

The shake-up has affected both customers and suppliers. The older market structure had depended on strict vertical integration both in developed and late industrialised countries, with multinational affiliates supplying the region (especially in Latin America). The steep investment required and the risks of opportunistic behaviour on the part of suppliers (who could take advantage of proprietary information unavailable to buyers) to secure and renew contracts for these large, incremental switching networks forged strong supplier-customer relationships. These vertical relations were both "formal", as in Italy and the United States, where the telecommunications carriers owned the equipment suppliers; and "informal", as in Japan, Germany and France, where local state carriers arranged long-term contracts with national suppliers (Fujitsu, Hitachi, NEC and Oki, Siemens and Alcatel respectively) and in Sweden and the United Kingdom, where state-owned carriers co-operated with local suppliers in the development of innovated equipment. In all of Europe both formal and informal ties were customary between equipment manufacturers and telecommunications service providers[2].

In Far East Asia and in most other late industrialising countries, demand for telecommunications services had been too small for such ties to form, at least until the 1970s. An important exception is Latin America, where affiliates of multinational firms that manufactured telecommunications equipment in Europe and the United States also supplied a large share of the telecommunications services[3].

By the end of the 1960s the international telecommunications industry was an oligopoly. Its solid vertical integration and the resulting density of multinational firms raised barriers to entry and mobility and slowed international trade.

The remarkable technological opportunities offered by the introduction of electronics in the 1970s, however, pushed the lethargic oligopoly into a new era of competition. This competition was shaped by three factors:

1. The R&D expenditure necessary to adopt the new electronic technology and develop digital telecommunications equipment was much higher than the R&D

expenditure for integrating incremental innovations into electromechanical systems, so the minimum efficient size required to recoup costs was substantial;

2. Economies of scale both technical and pecuniary (the larger the buyer, the lower the purchasing costs) and "learning economies" that reduce costs because of improved skills grew in importance, while fixed costs and the price of intermediate inputs like semiconductors and computer software were higher than with the old technology;

3. Discontinuity (incompatibility) between electromechanical and digital technologies negated any advantages that might have been gained from vertical integration, such as incremental innovations and improved growth.

Enhanced advantages of size owing to increasing economies of scale and learning processes, and the decline in advantages of vertical integration, had the combined effect of:

a) Raising barriers to new competitors while lowering mobility barriers within the telecommunications equipment industry itself and within the broader electronic sector[4];

b) Creating more opportunities for smaller customers who were not bound by formal vertical integration and so could buy more cheaply on competitive international markets.

As a consequence a fierce selection process began in the telecommunications industry, and rapid vertical disintegration and reshuffling of customer-producer relations took place[5].

The telecommunications industry, which for years had been one of the most stable multinational oligopolies[6], found itself at the end of the 1980s with only eight companies capable of manufacturing and installing electronic switching equipment. These firms, headquartered in only seven countries, possessed manufacturing affiliates in no less than thirty-five countries.

Table 11 indicates the evolution of international market shares of public switching equipment in the period 1982-1987. In these five years the cumulated market share of the top (then thirteen) firms increased from 88 per cent to 97 per cent. The number of top firms shrank to eight after mergers between Siemens of Germany and GTE of the United States; between ITT and Alcatel (United States and France); and between Plessey and GEC (both United Kingdom). The transition from electromechanical to electronic switching greatly diminished the number of independent firms and ran smaller manufacturers completely out of the international market. At the same time, however, lateral entry into regional markets was eased for new suppliers who had no previous vertical integration and already had a small foothold there.

At the end of 1987, eight different and basically incompatible models of digital switching technologies were competing on the international market (see Table 12), from the oldest, ESS-5, to the E 10, which was introduced in 1976 by French Alcatel and was especially suited for low density networks in developing countries. The last column of Table 12 is particularly interesting because it demonstrates the pressure that was put on suppliers, principally by late adopters, to sell switching equipment. It also suggests how badly companies like Italtel, GEC-Plessey, Alcatel and Siemens need to acquire bigger market shares before 1995, when the next "generation" of switching technology – based on optical fibres – is expected to appear.

Recent technological developments have enhanced the capacity of private branch switchboards (PBX) to such an extent that they can now serve as small switching centres in small local communities and in a rural environment.

The introduction of further innovations in the PBX technology in this direction is likely to offer important opportunities to the entry of new manufacturers of telecommunications switching equipment albeit limited to low telecommunications density markets. This seems to be so in the case of important attempts to extend the scope of the local manufacturing capacity recently developed in India and Brazil.

In India and Brazil the local telecommunications equipment industry, traditionally limited to the production of PBX, has recently made important investments to initiate the manufacturing of small switching centres. Such a process is likely to increase further the competition in the switching equipment markets especially in developing countries for two reasons.

First, a large share of their demand for switching equipment refers to low density telecommunications networks both because of low demographic density areas and because of low levels of penetration of telecommunications services.

Second, the access to such an upgraded PBX technology is easier than to the central switching one for many late industrialised countries where PBX manufacturing capacity had been already developed because of the incremental character of technological change with respect to the previous generations of PBX.

It seems however clear that barriers to entry will remain high in the markets for large central switches designed for large metropolitan areas and high density networks.

Switching equipment is one of the few industries in which international competition was so intense that it drastically reduced the number of firms: the demand for switching equipment by all countries could be supplied by these eight companies. On one hand, of course, such a system is antithetic to ideal conditions of market competition advocated by free traders. On the other hand, the concentration of fierce competition may favour, and indeed has favoured, the adoption of state-of-the-art technologies by countries that have little or no innovation capacity.

While supply was consolidating and the number of firms shrinking, old vertical integration ties were severed (particularly in the late industrialising countries of Latin America) or rearranged (as in Far East Asia)[7]. In the 1980s in Latin America, the Brazilian government (which in the 1970s had nationalised all interstate services) took over most of the private, often foreign-owned local carriers, whose numbers had dropped from 962 in 1972 to 103 in 1985. Spanish Telefonica, which specialises in services, had acquired 36 per cent of SESA in Brazil, and full control of Ecuador's telecommunications utility by the end of the 1980s. In Mexico and Chile, the government became the majority shareholder of the telecommunications service carriers[8].

Telefonos de Mexico, the state-owned carrier, digitalised the Mexican telecommunications infrastructure during the 1980s through a good bargaining strategy vis-à-vis local affiliates of multinational suppliers like Ericsson from Sweden and French Alcatel. Telefonos modernised the telecommunications network quickly and economically and was able to build a national production capacity for complementary inputs like software and printed circuits used in switching equipment[9].

In Singapore telecommunications services expanded rapidly during this period, surpassing all other key economic indicators like energy consumption, production of manufactured goods and GNP. This reflects a more than proportionate increase in demand for telecommunications and a strong supply effect. Telecoms (the Telecommunications Authority of Singapore) pursued an aggressive policy of technological upgrading – introducing advanced services like public data exchange, multi-access telex, teleconferencing, and electronic teletype terminals with memory – all based on digital switching acquired from Japanese companies in the early 1980s. Telecom's strategy has been a good example of a successful supply policy built on telematics and computer use with a subsequent fast rate of adoption of organisational innovations in the Singapore economy[10].

In the 1970s Indonesia was already one of the leaders in the introduction and use of satellites as structural components of an advanced telecommunications infrastructure. By 1984 Perumtel, the Public Telecommunications Corporation, was able to provide specialised data communications like packet switching to all Indonesian firms through the satellites Palapa B-1 and Palapa B-2. Over 100 earth stations were constructed throughout the Indonesian Islands for receiving signals from the Palapa system. By the mid-1980s an optic fibre cable had already been installed between Jakarta and Surabaya.

Indonesia has tightened state control over the telecommunications equipment and services industries, focusing on services development. According to a recent law (no. 3/1989), Perumtel has been named the sole manager of telecommunications services in the country. Its major responsibilities are introducing ISDN (Integrated Services and Data Network) and procuring all switching equipment[11,12].

International switching equipment trading in the 1980s was characterised by "vent for surplus" strategies by most exporters – selling at tomorrow's lower price, counting on future economies of scale[13]. With considerable learning advantages and prospects of long-term commitment from customers, suppliers are motivated to secure larger shares of the market and longer production schedules, and to do this they often sell their goods at prices lower than present average costs[14]. Most governments of industrialised countries helped spread this "dumping" practice with their international aid schemes of low-interest loans to developing countries. These countries were then tied to them for their digital switching equipment[15].

This kind of pricing behaviour on international markets could be advantageous in the short term for late-industrialising countries. Customers facing a competitive oligopoly of equipment suppliers might find some real bargains. In the long run, however, the incremental purchases and updating required for both the hardware and software of a telecommunications system that is physically incompatible with any other could force buyers into an expensive situation with few options[16].

Still, most late-industrialised countries in the 1980s had clear-cut advantages over developed countries; without obligation to buy from local "locked-in" suppliers whose prices were often higher than on international markets, they could concentrate on improving their national electronics industries and manufacturing capacity (as Mexico, Brazil and Singapore have done) through selective procurement strategies.

In sum, the pronounced advantages of late industrialising countries at the beginning of the 1980s emerged because of the following conditions:

1. Telecommunications equipment customers (carriers) were and basically still are large and monopolistic firms that, because of their "sewing-up" of the market, had strong bargaining power over suppliers;
2. The sudden decline of mobility barriers and the increase in economies of scale pushed telecommunications equipment firms operating in regional markets to

engage in heavy price competition. They could, of course, obtain large shares of the international market this way, and accelerate the shift from a multiregional oligopoly to a global market;

3. High levels of sunk costs in the establishment of new affiliates and the possibility of procuring long-lasting, irreversible commitments from new customers prompted the new global suppliers of digital equipment to make pre-emptive moves to force down prices. This was often possible through the economic and diplomatic support of home governments.

In many respects the 1980s were years of a "customer bonanza", at least for customers who could afford to take advantage of the international shopping opportunities, and thus drop import-substitution strategies. This has been easier for countries that had:

a) Low demand for telecommunications services at the end of the 1970s; and consequently

b) An obsolete or inefficient telecommunications infrastructure;

c) Low or non-existent telecommunications equipment manufacturing capacity;

d) Low national and international, formal and informal vertical integration with telecommunications equipment manufacturers;

e) Cultural proximity to aggressive, successful global players (such as North Africa with French Alcatel and Far East Asia with NEC and Fujitsu).

Far East countries fulfilled all these conditions. Latin American countries were somewhat hampered by early development and subsequent poor growth, with the stimulation of some manufacturing capacity and international vertical integration.

Many European countries, namely Germany, Italy, United Kingdom, and to some extent Japan, where the introduction of the fully digital system was long delayed, were heavy losers in this game[17]. Their efforts to protect their technological interests by harnessing captive markets to unsuccessful local manufacturers like Siemens, Italtel and GEC-Plessey, respectively, proved to be an increasingly vicious circle in terms of cost and technological progress.

NOTES AND REFERENCES

1. B. Lamborghini and C. Antonelli, "The Impact of Electronics on Industrial Structures and Firms' Strategies", in OECD, *Microelectronics, Productivity and Employment*, Paris, 1981.

2. See M. Fransman, *Beyond the Market: Cooperation and Competition in Information Technology in the Japanese System*, Cambridge University Press, Cambridge, 1990, for a detailed analysis of the Den Den system.

3. The United River Plate Telephone Co. Ltd., in Argentina; the Mexican Telephone and Telegraph Co.; SESA, part of the Brazilian network; and some of the Chilean network were owned by the US company IT&T. IT&T and the Swedish company Ericsson together owned the Argentinian Compania Argentina de Telefonos and the Compania Entrerriana de Telefonos. Canadian interests owned the Brazilian Telephone Company. See R. Katz, *The Information Society: An International Perspective*, Praeger, New York, 1988.

4. R.E. Caves and M.E. Porter, "From Entry Barriers to Mobility Barriers: Conjectural Decisions and Contrived Deterrence to New Competition", *Quarterly Journal of Economics* (May 1977), pp. 241-261.

5. A. Herrera, *La revolucion tecnologica y la telefonia argentina. De la Union Telefonica à la Telefonica Argentina*, Editorial Legaza, Buenos Aires, 1989.

6. N. Jecquier, "International Technology Transfer in the Telecommunications Industry", in D. Germidis, ed., *Transfer of Technology by Multinational Corporations*, OECD, Paris, 1977.

7. Affiliates of multinational companies specialised in the manufacture of electromechanical switching components. When these became obsolete, the affiliates either folded or diversified. Both supply and demand reduced all pre-existing "lock-in" effects.

8. D. Chudnovski, "La Industria de Equipos de Telecommunicaciones en la Argentina", CET-IPAL, Buenos Aires, 1985.

9. W. Peres, *Foreign Direct Investment and Industrial Development in Mexico*, OECD Development Centre, Paris, 1990.

10. Telecommunications Authority of Singapore, *Telecom's Annual Report*, Singapore 1982 and following years.

11. More than 45 per cent of total Indonesian investment in telecommunications equipment will be funded in coming years by World Bank – the equivalent of 600 000 line units out of a total 1 400 000 lines. The German company Siemens, a relative newcomer to the field, will benefit from a large share of this demand.

12. Dept. of Tourism, Post and Telecommunications, *Strategic Development Plan*, Jakarta, 1989.

13. R.E. Caves, "Vent-for-Surplus Models of Trade and Growth," in Essays in Honor of G. Heberler, *Trade Growth and the Balance of Payments*, Rand McNally, Chicago (1965), pp. 95-115.

14. The "lock-in" effects generated by standards incompatibility among public switching models lies at the core of the growing global oligopolistic rivalry. This rivalry is characterised by *"intertem-

poral price discrimination" based on the significant differences in price elasticity of demand, whether before or after the purchasing of a threshold share of total lines.

15. This has been well documented in many countries: France, Italy, Germany, Canada and Japan have supplied much telecommunications equipment in this way to newly industrialising countries.

16. Before the availability of the new switching standards, demand elasticity was much higher. Incompatibility of standards became an important building block for "lock-in" effects. See P.A. David, "Some New Standards for the Economics of Standardization in the Information Age", in P. Dasgupta and P. Stoneman, eds., *Economic Policy and Technological Performance,* Cambridge University Press, Cambridge, 1987.

17. H. Ungerer, *Telecommunications in Europe,* C.E.C., Brussels, 1988.

Table 11. **Breakdown of the shares of total public switching on international telecommunications markets 1982-1987**

	1982	1984	1986	1987
AT&T (USA)	25.5	26	28	26
NORTHERN TELECOM (Canada)	4.1	8	8	13
ITT (USA)	11.7	6	18	18
ALCATEL (France)	3.0	12		
GTE (USA)	7.0	6	13	13
SIEMENS (Germany)	7.0	11		
PHILIPS (Netherlands)	–	–	2*	–
ITALTEL (Italy)	2.5	2	2	2
ERICSSON (Sweden)	6.7	8	8	8
GEC (United Kingdom)	5.2	3	3	3
PLESSEY (United Kingdom)	1.7	3	4	3
FUJITSU (Japan)	1.0	2	2	3
NEC (Japan)	5.6	7	8	9
OTHERS	19.0	6	2	2
TOTAL	100	100	100	100

* In joint venture with AT&T.
Source: Interviews with ITALTEL managers.

Table 12. **R&D cost of digital switching systems ($ million) compared to public switching sales in 1987**

Company	Switch system	Year of Introduction	Estimated R&D Costs
ITT	System 12	1976	1 000
Ericsson	AXE	1977	500
CIT-Alcatel	E 10	1976	1 000
Northern Telecom	DMS	1977	700
GEC/Plessey/BT	System X	1983	1 400
Western Electric	ESS-5	1976	750
Siemens	EWS-D	1980	1 300
Italtel	UT	1982	500

	Sales in public switching	Required revenue (x)	Pay back time in years (y)
ITT	1 000	14 000	14
Ericsson	850	7 000	8
CIT-Alcatel	800	14 000	17
Northern Telecom	1 150	9 800	9
GEC/Plessey/BT	260	19 600	35
	300		
Western Electric	1 430	10 500	7
Siemens	1 100	16 000	14
Italtel	250	7 000	28

Source: Adapted from G. Dang Nguyen, "Telecommunications: A Challenge to the Old Order", in M. Harp, ed., *Europe and the New Technologies,* Frances Pinter, London, 1985; and A. Roobeck, "Telecommunications: An Industry in Transition", in H.W. de Jong, ed., *The Structure of the European Economy,* Kluwer Academic Publishers, Dordrecht and Boston, 1988.
(y) = Required sales/1987 sales in public switching.
(x) = Required sales revenues are estimated by Roobeck on the assumption that R&D should be equivalent to a maximum of 7 per cent of sales revenue. Roobeck expects each public switch when developed to earn four times its development costs in lifetime sales revenues. Expected lifetime for digital switches is conservatively estimated at 10 to 15 years.

Chapter 5

THE MACROECONOMICS OF DIFFUSION

The Evidence

Table 13 indicates that the growth rate of the overall telecommunications system was very fast in all countries listed in the period 1977-1987[1]. In some newly industrialising countries, however, the rate was extraordinary[2].

Canada and the United States increased the size of their total communications infrastructures, as measured by the amount of main lines per 100 inhabitants, by 20 per cent. In most European countries main line intensity increased by 50-60 per cent. It was much higher, though, in France (240 per cent), Ireland (193 per cent), Norway (185 per cent) and Sweden (121 per cent), and much lower in Switzerland (29 per cent).

In Latin America, growth rates for telecommunications were close to European levels. The best performances were recorded in the Far East, where South Korea multiplied its number of main lines by five, jumping from 4.2 per cent in 1977 to 20.7 per cent in 1987. Malaysia's lines increased from 1.8 per cent to 7.2 per cent. Turkey expanded lines by 350 per cent, from two lines to seven per 100 inhabitants.

In Table 13 we use the classifications Forging Ahead, Fast Growth and Falling Behind to measure the growth of main telecommunications lines per inhabitant in each country. The results are similar to those of Table 6, where we measured electronic switching diffusion in the categories of Forging Ahead, Fast Adopters and Falling Behind.

It can be seen from comparing the two tables that countries that "forged ahead" in the adoption of electronic switching from 1977 to 1987 were also those that forged ahead in the expansion of their telecommunications infrastructure during the same period[3].

In most developing countries both in Latin America and in the Far East, the number of lines installed during this period was much larger than the stock of electromechanical lines in existence when diffusion began in the mid-1970s. These countries thus had a remarkable opportunity to completely leapfrog the electromechanical technology, avoiding the expense of replacing obsolete (though young in age) capital stock and problems of technological cumulativity, and start their telecommunications infrastructure from scratch.

Investments have been proven to be central in this process of growth and modernisation. The combined pressures of an expanding telecommunications infrastructure and its systemic and discontinuous character incited many countries to vastly increase investment and benefit from steep declines in the prices of telecommunications equipment[4].

Table 14 presents data on countries (where available) indicating international distribution levels of gross investment in telecommunications from 1976 to 1986, in US dollars per capita[5,6]. We can see that investment efforts vary considerably from country to country. In

71

1986, for example, the ratio of the strongest investor to the weakest (Switzerland at $164.95 per capita to Thailand's $0.79) was extreme. Generally investment per capita in Latin American countries in 1986 was 30 to 40 times smaller than in the United States, Canada and most European countries, where investment that year averaged $75 per person. In the Far East, average 1986 levels were about $20 per capita, a third of the average of most OECD countries.

It is interesting to note that, in spite of the wide differences in investment levels noted in the first example between Switzerland and Thailand, by 1987 Thailand had achieved better diffusion of electronic switching, at 50.7 per cent of total switching capacity, than Switzerland at only 31.3 per cent. The data from Table 1, which show the rates of penetration of electronic switching in 1987, compared with data on rates of growth of investment in Table 14 suggest that a strong negative correlation takes place between the two: countries with high investment growth could reap the benefits of technological opportunities without "paying for" investment in the past.

Comparisons of data between these two tables also reveal that investment intensity corresponds closely to the increase in total lines per capita (i.e. the growth of demand) and penetration of electronic lines (i.e. the rate of diffusion).

In North America, where Canada and to a larger extent the United States moderately raised their investment levels during the 1976-1986 period, there was a slight increase in overall telecommunications intensity and diffusion of electronic switching.

In Latin America investment growth rates were much more lively, and during the same period investment per capita tripled, reflecting the fast growth of both the telecommunications infrastructure and diffusion of electronic switching.

European investment did not grow as quickly as in Latin America, but its rate of growth surpassed that of the United States, averaging around 200 per cent – though with two important exceptions. Germany expanded investments over 1976-1986 to 4.5 times its original outlay, with sluggish overall growth of its telecommunications infrastructure and low diffusion of electronic switching. Sweden multiplied its investments by 5.8, yet its telecommunications network, already at maximum levels by world standards, grew only marginally. Its diffusion of electronic switching, however, took place quite fast.

In France, Finland and Sweden, significant increases in investment accompanied good rates of diffusion. Low levels of investment in the United Kingdom corresponded to low rates of diffusion.

A vigorous rise in investment in Asia matched a swift diffusion process. South Korea multiplied its investments by a factor of 7.2, Sri Lanka by 6.7 and Hong Kong by 3.3. In these countries penetration of electronic switching quickly jumped to world standards. Some exceptions to this general trend, however, exist.

All in all the macroeconomic climate of growth of demand and the resulting investment seem to have been conducive to the brisk diffusion of electronic switching that occurred in the ten-year period between 1977 and 1987. Newly industrialising countries that had a rising demand for telecommunications services and a negligible pre-existing network could, with sufficient investment, adopt the new switching technology very rapidly. The data seem to offer distinct empirical evidence of a vital interaction between growth of demand and investment and diffusion of technological change.

Investments and Adoption: The Theory

The accepted theories on the diffusion of technological innovations are a logical outcome of microeconomic analysis, in which any consideration of the relationship between the diffusion of process innovations, adoption of innovated capital goods and investment decisions is almost completely absent. More specifically, the two standard microeconomic approaches to diffusion analysis, the "epidemic" and the "equilibrium," assume that diffusion is motivated by *a) exogenous factors* such as the technological performance and profitability of an innovated good as determined by its market price; and *b) a decision process or routine* based on either "bounded rationality" in the epidemic approach – rationality bounded by the limitations of risk, uncertainty, incomplete information about alternatives and complexity – or "hyper-rationality" in the equilibrium approach, which treats these limitations as minor and irrelevant[7].

Parallel to this microeconomic tradition stands a macroeconomic approach, which looks at diffusion in terms of the relationship between technological change and economic growth, and proposes that the capacity to adopt technological innovations is embedded in the process of accumulation and investment. Its main features are:

i) the endogenousness of the rate of diffusion; in other words, diffusion affected by variables that exist within the country or firm's economic system, not external factors; and

ii) the macroeconomic context of analysis.

Micro- and macroeconomic analyses have rarely converged. Nevertheless we will attempt to combine the two concepts for a broader understanding of international diffusion.

Technological change does not diffuse through the economic system only because of the technological advantages offered by innovated capital goods, as the standard microeconomic system holds, but because of a much more complex set of factors like *a)* the growth of demand, which affects *b)* investment behaviour; *c)* the profitability of adopting the innovation; *d)* relative factors costs; and *e)* the limited knowledge of managers and entrepreneurs[8].

Here we will try to show that an integrated approach to diffusion analysis can be built using elements of:

1. equilibrium models of diffusion;
2. epidemic models of adoption;
3. macroeconomic analysis of growth and investment.

The Equilibrium Models

An equilibrium approach to the analysis of diffusion has been developed in recent years to try to overcome the limits of demand-oriented epidemic models. The equilibrium approach is grounded exclusively on a microeconomic base and asserts that a delay in the adoption and use of a new technology is the result of the marketplace interaction of innovation suppliers and "hyper-rational" potential adopters[9].

From this point of view a heterogeneous group of potential adopters faces a dynamic environment in which factor costs and the technological characteristics of innovated capital goods are constantly changing. Each potential adopter rationally delays adoption until his or her operating costs are lower with the new technique than with the old[10].

Many different reasons for this process have been suggested:

a) Large firms usually pay higher wages than small firms, and so are motivated to adopt earlier. Smaller firms will adopt eventually, once their wage levels reach those of larger firms[11]. This leads to two hypotheses: i) diffusion levels should be higher in countries with larger firms and higher wages; and ii) diffusion should be faster when firms and wages grow faster[12];

b) Incremental technological change makes the innovated capital goods more attractive. In this case stronger profit potential can induce adoption even in firms with lower wages[13];

c) The eroding monopolies in the markets of innovated capital goods brought about by increased competition from "imitation" suppliers causes a fall in the goods' prices. This, of course, improves the profit potential for late adopters, and a wave of diffusion in new users takes place[14];

d) An increase in adoptions affords economies of scale in production. This means smaller production costs and consequent lower market prices of innovated capital goods[15];

e) Learning by doing over time helps reduce production costs and thus market prices, and this stimulates diffusion[16].

The equilibrium approach seems to suffer from two limitations. First, it rules out any of the effects of information asymmetries and limited knowledge on the decision-making process that are so important in the epidemic approach. Diffusion models built on epidemic learning may overemphasize the role of bounded rationality, but equilibrium models built on suppositions of hyper-rational, individualistic decision making leave no room for "learning through others" – and these are difficult assumptions to accept in a dynamic environment where decisions must be made without a known set of choices.

Second and more importantly, the equilibrium approach fails to take into account that a decision to adopt an innovated capital good means a decision to invest. As we have seen, investment and adoption decisions interact very closely.

The Epidemic Models

Criticism of the epidemic approach voiced by the equilibrium "school" has permitted a better evaluation of both its limits and strengths. Again, the standard epidemic model claims that diffusion rates are unequivocally determined by the epidemic "contagion" that early adopters spread to potential ones[17]. If we use Mansfield's original interpretation, contagion, or imitation, will improve the profit potential of adoption in a population characterised by bounded rationality and high transaction costs. Imitation can thus be viewed as an outcome of collective learning, so that the limited knowledge available only to first adopters gradually spreads to all potential adopters, and transaction costs decline as the stock of users grows.

This reassessment of the epidemic model does not exclude the complementary effects on adoption and diffusion of other factors such as changes in the innovation's profitability brought about by wage increases, the introduction of incremental innovations, or a decline in the prices of innovated capital goods[18,19].

A neo-epidemic model, on the other hand, incorporates into the effects of bounded rationality some insights of the equilibrium approach, especially the positive forces of supply and wages in diffusion. Still, the neo-epidemic model has the same restriction as the equilibrium one: it neglects to appreciate the crucial role of investments in adoption behaviour.

Growth, Investment and Diffusion

The pillar of a macroeconomic analysis of diffusion is the correlation between economic growth and diffusion, and more specifically the growth of demand for innovated capital goods that expands investment and consequently the purchasing opportunities for the goods.

In Mansfield's highly original work it was already possible to note that, even in an epidemic approach, elevated investment requirements for the adoption of an innovation can cause delays in adoption and slow diffusion rates[20]. Mansfield recognised that adopting an innovated capital good can be an act of "replacement investment" that displaces existing goods. Adoption may therefore be delayed by the age of the existing capital stock and by the amount of costs "sunk" in it[21].

Mansfield was also aware that the adoption of an innovated capital good, like any investment decision, may take place earlier when net investment is growing, and thereby avoid the pitfalls of replacement investment[22]. Mansfield and his followers, however, did not develop these ideas and the concept of net investment as both a propellor and (in the form of sunk costs) an obstacle to diffusion has fallen outside the traditional sphere of analysis.

In a different tradition of economic analysis the causal relationship between investment and diffusion receives full attention, but to such an extent that the possibility of non-adoption is overlooked, and diffusion is equated with capital accumulation plus investment[23]. In a post-Keynesian approach all new investment is expected to incorporate technological innovations (regardless of a firm or country's ability to adopt them) with an aim to improve labour productivity. Diffusion in this view is seen as the automatic outcome of investment.

The relationship between investment and labour productivity in Kaldor's well-known "technical change function" appears to be shaped by a vague perception of demand lags:

"Our 'TT' curve thus reflects not only 'inventiveness' in the strict sense, but the degree of technical dynamism of the economy in a broader sense – which includes not only the capacity to think of new ideas, but the readiness of those in charge of production to adopt new methods of production[24]".

Kaldor shapes the technical change function as a truncated logistic with an upward convexity that will flatten beyond a given point because of saturation effects. Kaldor does not provide any further explanation for such a shape that in a more traditionally diffusion-oriented context would be elaborated in more epidemic terms[25].

Salter develops Kaldor's ideas to make a distinction between "best-practice" technology and "average" technology (the newest, most performant innovated goods versus the average performance of existing ones), arguing that for given levels of technological change the rate of gross investment precisely determines the lag between observed productivity and best-practice productivity[26]. At the same time Salter distinguishes between replacement investment, net investment and gross investment. All innovated capital goods rely on gross investment. Its components, net and replacement investment, differ in many respects. Net investment is determined by new demand, and the prices that can be charged on the market for the product. Replacement investment is determined by comparing the profitability of innovated goods to the non-sunk costs (operating costs of material plus labour) of existing capital goods. When there is technical change after the introduction of innovated goods – and perfect competition – the purchase price of products declines. As soon as it becomes lower than the sunk costs of the existing plants, the firm will replace the old capital stock with the new[27].

It is clear that Salter has a theory on diffusion, though in a narrow context: an innovated capital good will be adopted only when a decision has been made for net investment, or when replacement-investment is not blocked by heavy sunk costs; i.e. when operating costs for the oldest part of the existing capital stock are higher than the prices of the product manufactured with the innovated capital goods, and when those prices are higher than the total unit cost possible with the innovated production process. Also, an innovated capital good will not be adopted at the same time by all potential users unless:

a) all potential adopters have the same levels of sunk costs; i.e. the same age and structure capacity of existing capital stock;

b) the prices of products manufactured with the innovated capital goods fall sharply below operating costs for all pre-existing plants, or operating costs for all existing plants surpass product prices.

In this view, diffusion, for given levels of growth and net investment, is determined by the amount of replacement investment that (for given levels of product price) is determined by the amount of sunk costs. Diffusion, then, would depend on:

1. Operating costs of existing capital goods[28];
2. Opportunities presented by technological change to purchase cheaper capital goods[29];
3. The different factor intensity (ratio of labour to capital) of the innovated production process and the possibilities it offers – for given levels of purchasing of innovated capital goods – to save labour and thus reduce total unit costs;
4. Costs for potential adopters of integrating the innovated capital goods into pre-existing capital stock.

The analytical steps used in Salter's model in the delay of a new technology's use are these[30]:

a) Innovated capital goods are assumed to be compatible with existing ones and so can be added on to form a "vintage" heterogeneous structure with differing age blocks of capital goods, differing labour productivity and physical characteristics[31];

b) The introduction of process innovations can expand industry output;

c) Process innovations engender a fall in the price of products;

d) In the long term, the fall in the price of new products motivates firms to replace older capital goods;

e) The reduction in total costs due to process innovations thus creates a surge of additional replacement investment.

In sum a careful interpretation of Salter's vintage model makes clear that both the neo-epidemic and equilibrium models suffer from a failure to consider the role of investment in adoption and the growing demand that stimulates it. Salter's model assumes that all investment is used to buy innovated capital goods, and that adoption delays are determined only by sunk costs and relative factor costs. Therefore he disregards the possibility that information asymmetries, limited knowledge and bounded rationality may force investors to keep buying older capital goods.

The Model

It seems evident by now that the diffusion of technological innovations embodied in capital goods must be considered the result of two distinct and related decisions: the one to invest, and the one to adopt.

With given levels of profitability of innovations, low levels of investment caused by dark market perspectives will make diffusion proceed slowly, even if all potential adopters could buy the innovation. Conversely, even when investment is strong because of good rates of demand, adoption will be slow if there is a lack of information about the innovation or high transaction costs.

Formally we can thus write:

[8] $Y = f(A, I)$

where Y = ratio of the stock of innovated capital goods to total installed capital
       A = adoption behaviour
       I = investment behaviour

We will now look at the determinants of adoption and investment behaviour.

a)   *Adoption behaviour*

Building on the epidemic tradition, it seems necessary to consider the effects of bounded rationality and limited knowledge in modelling the adoption process.

Potential adopters are not usually instantly aware of all process innovations being introduced. Transaction costs are higher for innovated capital goods than for traditional ones because of asymmetric information, opportunistic behaviour by suppliers who take advantage of this lack of information, and "small numbers exchange" – few suppliers and few buyers[32].

To model these effects we can rely on the traditional epidemic specification, which typically applies however more to the narrower case of adoption than diffusion:

[9] $A_t = f(Y_{t-1})^a$

where A_t = the ratio of adoptions of innovated capital good to purchases of old capital goods per unit of investment at time t;
       Y_{t-1} = the share of the stock of innovated capital goods already adopted at time t-1 compared to total stock of the capital goods used in the same production process.
       a = a parameter for learning effects.

Equation 9 tells us that adoption choices are influenced only by collective learning for a given level of profitability of the innovation to be adopted, or for the same innovation with the same technological and economic features in different industries, regions or countries[33]. It therefore seems to present the essence of adoption behaviour; in other words, the bounded rationality models in which adoption choices are the result of experience and collective learning[34].

b) *Investment behaviour*

Building on the analysis of the relationship between effective demand that spurs investment, factor costs and technological change as determinants in gross investment, we can model the investment function augmented by a technological change factor that takes into account both the price of the innovated good and the savings (in terms of operating costs) it makes possible. In this model technological change affects both net and replacement investment:

[10] $I_t = f[(O_{t-1}, K_t), TC]$

where I_t = investment levels
O_{t-1} = total industry output during year t-1
K_t = total capacity of the industry at the beginning of year t
TC = profitability of adoption

According to equation 10 we expect investments to be influenced by the growth of demand and the relative efficiency of the existing capacity[35].

If we substitute equations 9 and 10 into equation 8 we have a full model of diffusion in which investment interacts with the epidemic "spread" caused by collective learning:

[11] $Y = (Y_{t-1})^a [(O_{t-1}, K_t), TC]$

where Y = ratio of the stock of innovated goods to total installed capital
O_{t-1} = total industry output during year t-1
K_t = total capacity of the industry at the beginning of year t
TC = profitability of adoption
a = parameter for learning effects

Traditional theory holds that the profitability of adoption propels diffusion, as expected from the equilibrium models, but only when determinants such as low demand expectations do not prevent a long-term commitment of financial resources to the purchase of the innovated goods.

Not all investment, however, is put into innovated capital goods. Moreover the ability to choose an innovated capital good is influenced by more market signals than the purchase price: the spread of information and collective learning play a significant part.

The lagged diffusion levels here take into account both the role of learning in the preliminary phases of choosing an adoption, and of network externalities in terms of reductions in the costs of software and other complementary inputs and especially of skilled manpower.

A simpler, more straightforward version of our basic model can be devised by placing equation (9) into equation (8) to obtain:

[12] $Y_t = j [(Y_{t-1})^a, I]$

Our equation (12) can easily test the available data.

Investments and Diffusion: The Test

If we build on the theoretical considerations previously outlined and other evidence we have seen so far, we can econometrically test the intensity and effectiveness of the link between investments and diffusion.

According to the basic model shown in equation 12 we expect to prove that:

1. Diffusion results from investment as well as adoption behaviour. Without investment, diffusion – the purchase of innovated goods – cannot take place. We therefore expect to verify that the flow of investments exerts strong effects on the diffusion process;
2. Adoption is made on the basis of imitation, through *cognitive externalities* (collective learning) and learning by doing, and *network externalities* like declining costs and increased availability of complementary inputs, maintenance and skills qualifications. Both these cognitive and network externalities are generated by the stocks of innovated machines and become effective after a high threshold of penetration;
3. Diffusion of electronic switching is strongly influenced by the ratio of the flow of new capacities, as determined by investments, to the installed base of electromechanical switching.

Consequently we expect to test empirically that:

a) A transition to electronic switching is slower in industrialised countries with recent vintages of electromechanical exchanges built with large excess capacity; adding to this capacity is cheaper than buying electronic switching systems;
b) More investment is necessary for early adopters because of the higher cost of integrating electronic switching into a pre-existing telecommunications infrastructure;
c) Investment is more effective in countries with low levels of telecommunications density at the beginning of the diffusion process. In these countries, in fact, net investment suffices to reach good penetration levels. Countries with high telecommunications density, on the other hand, experience a long delay and repeated investment before electronic lines make up a significant share of total switching capacity.

A direct estimate of differing effects of investment among countries can be obtained by testing our simple model. We will use equation 12 to test the intensity of the relationship between investment and the penetration of electronic lines on total switching capacity. Our model and the regression equation is shown as follows:

$$[13] \quad Y(t) = a + b_1 \, Y(t-1) + b_2 \, INV \, (t-n)$$

Specifications of the variables used in equation 13 are given below:

$Y(t)$ = penetration levels of electronic switching in total switching capacity in each country, as measured by the percentage ratio of electronic lines to total lines installed in the period 1977-1986;

$INV(t-n)$ = investment effort in each country as measured by the ratio of total gross investments in telecommunications (excluding land and building), converted in US$ to total population in the years 1974-1986.

79

All variables are transformed into natural logarithms to express the multiplicative relationship between adoption behaviour and investment behaviour (influenced by the penetration levels already reached) and their effect on the diffusion process[36].

The tests have been performed with a one, two and three year time lag of INV (Investments per capita). Results of the best estimates with the different time lags are given in Table 15. Within the limits of the short time span available that reduces the observations to 11, further reduced to 10 by Y (t-1), results can be considered a strong confirmation of the hypotheses we have presented. In all countries used, the explanatory power of equation 13 is very high. Total variance explained is below 80 per cent only in Costa Rica, Colombia and Sri Lanka.

The autoregressive character of the adaptive behaviour of adoption specified in equation 13 properly serves our test. The parameters estimated of Y (t-1) are significant and have strong values in most cases. The distribution of estimated values closely parallels that of diffusion rates (estimated by equation 1 in Chapter 3), and can itself be a measure of the effects of imitation on diffusion. Investment exerts a powerful effect in most countries. As expected the elasticity of diffusion to investment is especially low in countries where diffusion started very early, where investment growth was slow, and where telecommunications density was higher.

Our interpretation of these results is this: every dollar per capita of increase of investment has had a strong effect on diffusion in countries that were late starters and fast adopters and a weak effect in countries that started early. Latecomers had an edge on early starters for two reasons:

- First, purchasing costs of electronic switching hardware declined substantially from an average on international markets of $600 per line in 1976 to $400 in 1980/1981, and from $250 in 1985 to $189 in 1987. Latecomers could buy many more lines with less dollars. Telecommunications equipment has undergone a steep decline in real and monetary prices, while general price levels are increasing.
- Second, the cost advantages of electronic lines over electromechanical ones were lower for countries with high levels of pre-existing telecommunications infrastructure because of higher costs of the gateways necessary to accommodate electronic switching into the pre-existing electromechanical network.

In conclusion, we can say that low telecommunications density countries could achieve high diffusion performance because of two factors: a) the rising growth of demand for telecommunications and the corresponding strong investment that allows large amounts of innovated lines in countries with b) very small stocks of obsolete switching capacity.

The interaction between investment efforts, epidemic adoption behaviours and the diffusion process seems confirmed, and offers some interesting possibilities for leapfrogging for latecomer countries.

NOTES AND REFERENCES

1. Data for telecommunications lines have been gathered from ITU (International Telecommunications Union), *Yearbook of Common Carriers, Telecommunications Statistics*, Geneva, 1973 and following years.

2. These figures are especially noteworthy when considering that in these years many countries scrapped electromechanical switching capacity and installed analog and more recently digital switching. In some countries because of intense scrapping total telephone density declines.

3. The rank correlation is 0.67.

4. As already mentioned, the price fell from $600 to $189 per line in the period 1975-1987.

5. Data consist of gross investments in telecommunications excluding building and land. The share of resources specifically invested in buying and installing new switching capacity usually is between 35 per cent and 45 per cent of the total. It seems appropriate, however, to analyse the distribution of gross investment because of the effort involved in the modernisation process (including transmission) imposed by the adoption of electronic switching.

6. Data for investments (in US$) come from Siemens, *International Telephone Statistics*, Munich, 1984 and following years; ITALTEL, *L'annuario delle telecomunicazioni*, Milan, 1984 and following years, and ITU (International Telecommunications Union), *Yearbook of Common Carriers, Telecommunications Statistics*, Geneva, 1973 and following years.

7. According to H.A. Simon, rationality is bounded by risk, uncertainty, incomplete information about alternatives, complexity, cost of computational efforts. Bounded rationality substitutes satisfying behaviours for maximizing ones. Conversely, hyper or global rationality considers these limitations irrelevant or minor. See *Models of Bounded Rationality: Behavioral Economics and Business Organization*, MIT Press, Cambridge, 1982.

8. "It is of little help in comparative growth analysis to try and explain why growth rates differ while at the same time assuming that knowledge is a free good that falls like manna from heaven. Technological knowledge, technological progress and productivity growth are very much the result of a concerted effort on the part of the economic actors involved to alter technologies and are best viewed as something that is to a large extent endogenous. Second, the importance of investment in its role of adopting and modifying new technologies and the importance of the latter in explaining why rates of growth in manufacturing output differ across countries and over time cannot be overemphasized. Third, it is important that there be an entrepreneurial or managerial class oriented to exploiting the available stock of technology if rapid growth is to result. Fourth, while nothing has been said about the determinants of the investment ratio, the notion that the investment ratio is something determined by the desires of entrepreneurs is certainly consistent with the analysis just presented. In contrast, in neoclassical analysis, the rate of investment is determined by the propensity to save and not the other way around", J. Cornwall, *Modern Capitalism: Its Growth and Transformation*, Martin Robertson, London (1977), p. 121.

9. "The diffusion of innovations is not simply a temporary disequilibrium phenomenon reflecting differences in the alacrity with which different entrepreneurs respond to a uniform economic stimulus, the opportunity for each to make a profit by cutting costs. Rather diffusion is portrayed

81

as the reflection of a changing (equilibrium) distribution among the different techniques, each one chosen rationally by a member of a heterogeneous population of firms, a population for which it could not be said that the latest method that has become available at any moment *ipso facto* constituted the dominant, best practice technique." P.A. David, *Technical Choice, Innovation and Economic Growth,* Cambridge University Press, Cambridge, 1975.

10. Formally in the equilibrium approach adoption will take place when:

[1] $W (L_O - L_N) > P_N - P_O$

where W = wage rate and per unit of production;
    L_O = labour requirements of the old methods;
    L_N = labour requirements of the new technology;
    P_N = capital cost of the new technology; and
    P_O = capital cost of the old technology.

It is thus clear that over time, adoptions Y will increase for all increases of W and of (L_O-L_N) and decreases of $(P_N - P_O)$.

[2] $Y = f [W, (L_N - L_O), (P_N - P_O)]$

11. S. Davies, *The Diffusion of Process Innovations,* Cambridge University Press, London, 1979.

12. P.A. David, *Technical Choice, Innovation and Economic Growth,* Cambridge University Press, Cambridge, 1975.

13. P. Stoneman and N. J. Ireland, "The Role of Supply Factors in the Diffusion of New Process Technology", *Economic Journal* (March 1983), pp. 66-78.

14. C. Antonelli and L. Ghezzi, "Un'analisi teorica dei processi di diffusione dell'innovazione tecnologica in regime di monopolio temporaneo", *Giornale degli Economisti e Annali di Economia 3* (1987), pp. 125-148.

15. J.S. Metcalfe, "Impulse and Diffusion in the Study of Technical Change", *Futures* (October 1981), pp. 347-359.

16. C. Antonelli and G. Gottardi, "Interazioni tra produttori e utilizzatori nei processi di diffusione tecnologica", *L'Industria* 4 (1988), pp. 599-636.

17. Formally the epidemic model assumes that:

[3] $\dot{Y} = b_1 \dfrac{Y (N - Y)}{N} \dfrac{1}{t}$

where Y = adoption levels; N = ceiling number of potential adopters; and t = time. Therefore, the rate of increase of adoption is a function of the product of the numbers of "uninfected" members of a fixed population and the share of that population already "infected".

18. C. Antonelli, "Profitability and Imitation in the Diffusion of Process Innovations", *Rivista Internazionale di Scienze Economiche e Commerciali,* (February 1990), pp. 109-126.

Antonelli, "The Role of Technological Expectations in a Mixed Model of International Diffusion of Process Innovations: The Case of Open-End Spinning Rotors", *Research Policy 5* (1989), pp. 273-288.

19. Formally such a neo-epidemic model thus argues that:

[4] $\dot{Y} = \dfrac{b_1}{t} \cdot \dfrac{N (N-Y)}{N} + \dfrac{b_2}{S} \dot{S} \dfrac{Y (N-Y)}{N}$

where Y = adoption levels; N = ceiling number of potential adopters; t = time; and S = some relevant aspects of supply behaviour (such as prices and/or performances of innovated capital goods) or wage levels. Diffusion therefore is assumed to result from multiplicative interaction of *time,* which expresses the classical process of imitation, the effects of wage levels and supply forces (such as the introduction of incremental technological innovations) and a decline in the market prices of innovated capital goods.

20. "Third, for equally profitable innovations, $L_{ij}t$ [i.e. the proportion of 'hold-out' – firms not using the innovation – at time t] should tend to be smaller for those requiring relatively large investments. One would expect this on the grounds that firms tend to be more cautious before committing themselves to such projects and that they often have more difficulty in financing them. According to the interviews, this factor is often important", E. Mansfield, *Industrial Research and Technological Innovation,* W.W. Norton, New York, 1968, p. 138.

21. "First, one might expect $L_{ij}(t)$ to be smaller if the innovation replaces equipment that is very durable. In such cases there is a good chance that a firm's old equipment has a relatively long useful life, according to past estimates. Although rational economic calculation might indicate that replacement would be profitable, firms may be reluctant to scrap equipment that is not fully written off and that will continue to serve for many years...." E. Mansfield, *Industrial Research and Technological Innovation,* W.W. Norton, New York, 1968, p. 146.

22. "Second, one might expect $L_{ij}(t)$ to be higher if firms are expanding at a rapid rate. If they are convinced of its superiority, the innovation will be introduced in the new plant built to accommodate the growth in the market. If there is little or no expansion, its introduction must often wait until the firms decide to replace existing equipment...." E. Mansfield, *Industrial Research and Technological Innovation,* W.W. Norton, New York, 1968, p. 147.

23. "Hence the speed at which society can 'absorb' capital (i.e. it can increase its stock of man-made equipment, relatively to labour) depends on its technical dynamism, its ability to invent and introduce new techniques of production. A society, when technical change and adaptation proceed slowly, when producers are reluctant to abandon traditional methods and to adopt new techniques is necessarily one where the rate of capital accumulation is small. The converse of this proposition is also true: the rate at which a society can absorb and exploit new techniques is limited by its ability to accumulate capital", N. Kaldor, "A Model of Economic Growth", *Economic Journal,* (December 1957), p. 595.

24. N. Kaldor, "A Model of Economic Growth", *Economic Journal,* (December 1957), p. 596.

25. "The shape and the position of the [TT] curve reflect both the magnitude and the character of technical progress as well as the increasing organisational, etc. difficulties imposed by the faster rates of technical change...there is likely to be some maximum beyond which the rate of growth of productivity could not be raised...." N. Kaldor, "A Model of Economic Growth", *Economic Journal,* (December 1957), p. 596.

26. "A second feature of the model is the role of gross investment as the vehicle of technical change. When there is no technical change, investment is required only to make good the depletion of existing capital stock through physical deterioration, and to add to this stock. But when technical change is taking place, gross investment has another extremely important role: that of providing the necessary specialised capital equipment required for new techniques, irrespective of whether or not they are more or less mechanised than their predecessors. Without gross investment, improving technology that requires new capital equipment simply represents a potential for higher productivity; to realise this potential requires gross investment. An economy with a low rate of gross investment is restricted at the rate of which new techniques can be brought into use: an economy with a high rate of gross investment can quickly bring new methods into use and thus realise the benefits of improving technology. In this way, the rate of gross investment is a vital determinant of the extent to which observed productivity lags behind best-practice technology", W.E.G. Salter, *Productivity and Technical Change,* Cambridge University Press, London, 1960, p. 63.

27. "This single model, which provides a basic framework for the analysis of productivity movements, is based on two well-known principles. The first is that capital equipment in existence earns rents in a manner analogous to land. For this reason, the immediate general adoption of new techniques which require investment is uneconomic since new plants will only be constructed when receipts are sufficient to cover all outlays, while existing plants will remain in operation so long as they earn a positive rent, even though their productivity is lower and their operating costs higher than a modern plant. The second principle is that to employ all new techniques, irrespective of

whether they are more or less mechanised than their predecessor, requires an investment outlay. Consequently, gross investment is the vehicle of new techniques, and the rate of such investment determines how rapidly new techniques are brought into general use and are effective in raising productivity", W.E.G. Salter, *Productivity and Technical Change*, Cambridge University Press, London, 1960, p. 65.

28. "The range of techniques in existence is defined by the condition that plants are not scrapped until their operating costs (labour plus materials) per unit of output equal price; or by the condition that replacement will not be profitable until their operating costs equal total costs (including amortization and interest) of a new plant", W.E.G. Salter, *Productivity and Technical Change*, Cambridge University Press, London, 1960, p. 65.

29. "In summary we may think of technical progress as raising the productivity of labour in three distinct but interrelated steps: i) the direct effect of technical advances, ii) increased mechanisation induced by cheaper capital goods, and iii) higher standard of obsolescence also induced by cheaper capital goods", W.E.G. Salter, *Productivity and Technical Change*, Cambridge University Press, London, 1960, p. 73.

30. It seems possible, assuming conditions of perfect competition, to formalize Salter's model with a system of three equations:

   [5] $Y = f(\overset{+}{I}_R, \overset{+}{I}_N)$

   where Y = diffusion, I_R = replacement investments, and I_N = net investment.

   [6] $I_R = f(\overset{-}{L}_N, \overset{+}{L}_O, \overset{+}{P}_N, \overset{+}{P}_O, \overset{+}{W})$

   [7] $I_N = (GNP_t - GNP_{t-1})$

   where L_N = labour requirements of the new technology per unit of production
   L_O = labour requirements of the old methods per unit of production
   P_N = capital cost of the new technology per unit of production
   P_O = scrap value of the old capital per unit of production
   W = wage rate and other operating costs.
   GNP = equilibrium level of output

31. "This assumption is in fact of great relevance. The degree of physical compatibility of different vintages of capital goods embodying different levels of technological change has major effects on the economic effects of sunk costs. When piecemeal addition of new vintages is possible indivisibilities are reduced and firms can fund new vintages, i.e. make replacement investments which embody technological innovations on a marginal basis. When physical compatibility is low however and addition of new vintage is difficult or expensive the relevance of sunk costs increases as well as the delay in replacement investment", M. Frankel, "Obsolescence and Technical Change in a Maturing Economy", *American Economic Review* (June 1955), pp. 296-319.

32. See the critique of R. Nelson to the standard vintage models: "But such models require perfect knowledge and foresight on the part of firm managers, and perfect knowledge and mobility on the part of factors"; see "A Diffusion Model of International Productivity Differences in Manufacturing Industry", *American Economic Review* (December 1968), pp. 1219-1248.

33. If we assume that the relative profitability of adoption (P) of a capital good incorporating technological innovations changes over time and across countries and industries because of the dynamics of factor costs, the behaviour of supply forces, and the introduction of incremental innovations, we can make explicit its multiplicative role:

   [9] $A = f(\overset{+}{Y}_{t-1}, \overset{+}{P})^a$
   A = the ratio of adoptions of the innovated capital good with respect to purchase of old capital goods per unit of investment at time t;
   $\overset{+}{Y}_{t-1}$ = the stock of innovated capital goods already adopted at time t-1;
   P = the profitability of adopting an innovated capital good instead of a traditional one.
   a = a parameter for learning effects

34. "According to the theory we are talking about people modify their behaviour not only in response to changing external (market) conditions, but also in response to what they have done and what has happened to them in the past. This would seem to be an integral part of the human condition. It would hardly require discussion if it were not for the fact that such a dependence is usually ignored by economists", R.H. Day, "Endogenous Preferences and Adaptive Economizing", R.H. Day and G. Eliasson, eds., *The Dynamics of Market Economics,* North Holland, Amsterdam, 1986, p. 155.

35. One can also put forward the hypothesis that technological change affects the absolute levels of desired capacity, thus enhancing the levels of net investment. Technological change is thus expected to generate increases of total demand via a decrease in the absolute propensity to save and a reduction in price of existing products. Demand elasticity is thus expected to be > 1.

36. A major advantage of the logarithm specification of equation 13 is that it can be easily interpreted as the result of the following passage:

[14] $\log Y_t - \log Y_{(t-1)} = a + b_1 \log INV$

Equation 13 thus tests the effects of investment efforts on the growth of penetration of innovated capital goods and at the same time makes it possible to measure the imitation effects generated by penetration levels already reached.

Table 13. **An international classification according to the rates of growth of telecommunications intensity (1977-1987)**

Latin America	OECD	Far East Asia	Other asian & african countries
FORGING AHEAD >200%			
Brazil	France Yugoslavia Turkey	S. Korea Thailand Singapore Malaysia Indonesia Sri Lanka	Pakistan Syria Malta Algeria Tunisia
FAST GROWTH >140 <199.9%			
Colombia Chile Ecuador Venezuela Costa Rica Mexico Argentina Uruguay	Australia Austria Belgium Finland Germany Greece Ireland Italy Norway Spain United Kingdom Denmark Netherlands	Hong Kong	Algeria S. Africa Israel UAE
FALLING BEHIND <139.9%			
	Sweden Switzerland New Zealand Japan USA	Philippines	Iran

Table 14. Total annual gross investments in telecommunications

In US$ per inhabitant

	1976	1977	1978	1979	1980	1981	1982	1983	1984	1985	1986
Canada	80.03	76.04	73.30	79.58	88.29	98.92	74.42	80.48	77.15	75.48	80.23
USA	52.21	57.71	68.02	77.14	82.48	84.70	80.11	68.78	67.90	87.88	87.76
Costa Rica	10.13	10.62	12.95	7.11	11.52	9.82	14.28	5.99	5.33	6.77	9.81
Mexico	3.89	4.91	6.03	4.92	6.29	8.03	7.07	8.57	16.33	16.91	16.46
Colombia	0.93	0.95	0.89	1.96	2.85	5.55	4.93	2.24	5.09	2.46	2.83
Finland	47.11	44.76	47.49	49.96	51.82	52.65	50.79	48.24	45.50	68.16	80.34
France	51.05	65.65	86.71	98.58	88.75	67.85	60.18	54.62	49.76	76.59	103.92
Germany	27.91	4O.18	57.21	72.15	77.44	75.71	76.61	66.73	67.10	98.03	125.59
Italy	31.20	36.73	42.06	44.56	45.87	40.57	47.19	47.19	45.23	57.62	77.08
Spain	22.00	24.26	34.20	36.91	38.52	35.35	29.28	25.38	25.21	29.80	38.42
Sweden	25.62	22.55	27.65	36.29	48.12	75.49	65.52	65.67	59.86	74.92	149.63
Switzerland	70.94	85.66	96.28	94.49	90.97	104.95	97.37	98.46	83.67	115.51	164.95
United Kingdom	26.61	27.14	24.64	33.67	43.60	56.89	39.57	37.63	33.32	35.46	43.13
Malaysia	9.10	9.78	11.68	12.96	19.77	20.10	21.67	28.77	40.43	42.98	30.73
Philippines	9.19	9.90	8.94	13.80	17.19	24.31	24.79	29.68	28.65	29.48	31.12
Singapore	21.74	27.51	26.92	27.25	37.30	40.55	43.24	47.31	53.55	54.58	44.37
Thailand	0.45	0.50	0.58	0.67	0.73	0.71	0.77	0.78	0.73	0.70	0.79
Hong Kong	4.00	4.06	9.76	12.84	15.88	16.81	18.52	12.40	14.34	13.13	14.51
Japan	36.46	41.34	56.27	59.73	50.49	60.49	61.84	52.76	54.45	52.58	58.39
S. Korea	5.53	7.30	10.05	13.80	16.11	22.87	27.31	33.30	29.46	31.19	39.78
Sri Lanka	0.19	0.27	0.30	0.43	0.76	0.74	0.98	1.03	1.66	1.62	1.27

Source: Siemens, Italtel, I.T.U. Data consist of gross investments in telecommunications excluding building and land. The share of resources specifically invested in buying and installing new switching capacity usually is between 35 per cent and 45 per cent of the total. It seems appropriate, however, to analyse the distribution of gross investment because of the effort involved in the modernisation process (including transmission) imposed by the adoption of electronic switching.

Table 15. Results of GLS estimates of equation 13

Canada	log Y =	2.537 (2.094)	+	0.873 (9.093)	log Y(t-1)	+	0.782 (1.754)	log INV(t-3)	$\bar{R}^2 = 0.928$
USA	log Y =	0.867 (1.353)	+	0.738 (12.536)	log Y(t-1)	−	0.105 (0.632)	log INV(t-3)	$\bar{R}^2 = 0.985$
Costa Rica	log Y =	32.597 (1.722)	+	1.493 (3.559)	log Y(t-1)	+	6.470 (1.696)	log INV(t-3)	$\bar{R}^2 = 0.743$
Mexco	log Y =	6.581 (3.665)	+	0.285 (1.020)	log Y(t-1)	+	1.142 (3.602)	log INV(t-2)	$\bar{R}^2 = 0.874$
Colombia	log Y =	8.066 (1.384)	+	0.400 (1.540)	log Y(t-1)	+	1.069 (1.192)	log INV(t-3)	$\bar{R}^2 = 0.714$
Finland	log Y =	37.673 (3.022)	+	0.928 (11.137)	log Y(t-1)	+	12.205 (2.961)	log INV(t-3)	$\bar{R}^2 = 0.960$
France	log Y =	1.380 (2.723)	+	0.852 (23.068)	log Y(t-1)	+	0.228 (1.225)	log INV(t-2)	$\bar{R}^2 = 0.987$
Germany	log Y =	9.446 (2.115)	+	0.045 (0.142)	log Y(t-1)	+	3.577 (2.083)	log INV(t-2)	$\bar{R}^2 = 0.828$
Italy	log Y =	17.658 (1.602)	+	0.810 (4.417)	log Y(t-1)	+	5.435 (1.546)	log INV(t-3)	$\bar{R}^2 = 0.858$
Spain	log Y =	8.354 (1.326)	+	0.729 (5.928)	log Y(t-1)	+	2.187 (1.205)	log INV(t-1)	$\bar{R}^2 = 0.843$
Sweden	log Y =	5.320 (2.432)	+	0.604 (3.434)	log Y(t-1)	+	1.606 (2.167)	log INV(t-1)	$\bar{R}^2 = 0.911$
Switzerland	log Y =	18.346 (1.671)	+	0.892 (4.726)	log Y(t-1)	+	7.345 (1.611)	log INV(t-3)	$\bar{R}^2 = 0.808$
United Kingdom	log Y =	3.385 (1.825)	+	0.541 (2.811)	log Y(t-1)	+	0.524 (1.168)	log INV(t-3)	$\bar{R}^2 = 0.678$
Malaysia	log Y =	3.971 (5.485)	+	0.372 (3.780)	log Y(t-1)	+	0.427 (4.070)	log INV(t-2)	$\bar{R}^2 = 0.979$
Philippines	log Y =	1.526 (0.476)	+	0.746 (4.073)	log Y(t-1)	+	0.124 (0.178)	log INV(t-3)	$\bar{R}^2 = 0.954$
Singapore	log Y =	4.242 (2.684)	+	0.503 (3.373)	log Y(t-1)	+	0.687 (2.096)	log INV(t-3)	$\bar{R}^2 = 0.955$
Thailand	log Y =	53.881 (2.343)	+	0.662 (3.627)	log Y(t-1)	+	7.269 (2.310)	log INV(t-3)	$\bar{R}^2 = 0.896$
Hong Kong	log Y =	1.731 (3.374)	+	0.785 (16.179)	log Y(t-1)	+	0.207 (2.067)	log INV(t)	$\bar{R}^2 = 0.985$
Japan	log Y =	17.113 (1.372)	+	0.761 (3.309)	log Y(t-1)	+	5.620 (1.331)	log INV(t-3)	$\bar{R}^2 = 0.730$
S. Korea	log Y =	24.096 (2.312)	+	0.171 (0.522)	log Y(t-1)	+	6.081 (2.218)	log INV(t)	$\bar{R}^2 = 0.922$
Sri Lanka	log Y =	24.211 (1.220)	+	0.124 (0.284)	log Y(t-1)	+	3.093 (1.137)	log INV(t)	$\bar{R}^2 = 0.634$

Table 16. Telephone main lines per 100 inhabitants 1977-1987

	1977	1978	1979	1980	1981	1982	1983	1984	1985	1986	1987	Rate of growth
Canada	38.14	39.30	40.30	41.43	42.13	41.70	41.84	45.14	46.34	48.06	44.50	116.6
USA	38.57	39.50	40.30	41.20	40.97	40.91	45.35	43.72	45.06	47.00	48.10	124.7
Costa Rica	5.20	5.87	6.20	7.10	7.29	7.84	7.88	8.06	8.20	8.05	8.60	165.4
Mexico	3.32	3.56	3.73	3.73	4.14	4.25	4.64	4.67	4.53	4.80	4.90	147.6
Argentina	6.50	6.55	6.64	6.74	7.76	8.17	7.61	8.64	8.87	9.21	9.60	147.7
Chile	3.02	3.19	3.16	3.27	3.38	3.58	3.74	4.05	4.41	4.49	4.90	162.2
Colombia	3.74	3.93	3.87	3.99	4.01	4.48	4.43	5.04	5.69	6.00	7.20	192.5
Ecuador	2.45	2.56	2.69	2.72	2.80	3.22	3.29	3.03	3.01	3.17	4.40	179.6
Venezuela	4.81	4.98	5.15	5.40	5.80	6.50	6.28	6.51	6.96	7.43	9.20	192.3
Brazil	2.71	3.16	3.29	3.93	4.16	4.47	4.73	4.96	5.12	5.24	5.60	206.6
Uruguay	7.21	7.28	7.49	7.59	7.74	8.00	8.32	8.68	9.83	10.47	10.60	147.0
Algeria	0.94	1.12	1.37	1.60	1.83	2.02	2.13	2.16	2.42	2.52	2.71	288.3
Morocco	0.75	0.79	0.80	0.83	0.91	0.94	0.96	1.00	1.09	1.11	1.13	150.0
S. Africa	4.97	5.88	6.19	6.87	7.17	7.89	8.00	8.16	8.42	8.36	8.40	169.0
Tunisia	1.35	1.46	1.64	1.78	1.85	2.05	2.18	2.42	2.60	2.90	3.00	222.0
Austria	23.24	25.11	27.10	29.19	30.75	32.25	33.75	34.97	36.11	37.25	38.40	165.2
Belgium	20.84	22.13	23.51	24.97	26.44	27.82	28.84	29.94	30.99	32.99	33.90	163.0
Denmark	38.37	40.20	42.07	43.46	44.72	45.95	46.97	48.25	49.72	52.35	55.10	143.5
Finland	31.48	32.81	34.51	36.34	38.16	39.83	41.50	42.94	44.59	46.11	47.90	152.1
France	18.49	22.01	25.48	28.89	32.09	35.04	37.39	39.26	40.77	42.21	44.70	241.2
Germany	25.65	22.82	30.79	33.39	35.28	36.87	38.26	40.27	41.93	43.24	45.00	175.4
Greece	20.59	21.61	22.74	23.45	24.55	25.81	27.49	29.48	31.32	33.03	34.70	168.5
Ireland	11.67	12.28	12.80	13.93	15.46	16.56	17.49	18.93	19.79	21.23	22.50	192.8
Italy	19.04	20.15	21.35	22.77	24.63	25.89	27.39	28.93	30.40	31.85	33.30	175.0
Netherlands	28.30	30.60	32.68	34.43	35.72	36.88	37.94	39.05	40.08	41.25	42.40	149.8
Norway	25.02	25.93	27.31	29.25	31.60	34.58	37.59	39.98	42.06	44.55	46.40	185.4
Spain	15.59	16.82	18.01	19.32	20.31	21.02	21.86	23.14	23.76	25.20	26.20	168.0
Sweden	54.37	55.47	56.34	57.95	58.81	59.63	60.22	61.46	62.72	64.10	65.60	120.6
Switzerland	41.08	42.24	43.34	44.46	45.50	46.54	47.75	48.95	50.15	51.45	52.90	128.7
United Kingdom	27.17	29.48	31.67	33.11	34.08	34.75	35.82	37.07	38.28	39.00	42.40	156.1
Yugoslavia	4.95	5.50	6.10	6.74	7.47	8.24	9.05	9.86	10.50	11.65	12.90	260.1
Malta	11.08	13.79	14.90	16.38	18.98	20.33	23.60	24.08	24.66	28.30	28.50	257.2
Australia	27.76	29.24	30.92	32.51	35.85	35.87	37.21	38.81	40.13	41.44	41.90	150.9
Indonesia	0.17	0.19	0.22	0.25	0.28	0.31	n.a.	n.a.	n.a.	0.38	0.40	235.3
Malaysia	1.76	2.03	2.40	2.95	3.64	4.27	4.76	5.55	6.11	6.44	7.20	409.1
New Zealand	34.10	34.72	35.06	36.10	36.78	37.08	37.53	38.28	39.37	39.98	41.60	121.9

Table 16. Telephone main lines per 100 inhabitants 1977-1987 (Cont')

	1977	1978	1979	1980	1981	1982	1983	1984	1985	1986	1987	Rate of growth
Philippines	0.82	0.81	0.83	0.87	0.90	0.96	0.85	0.90	0.88	0.91	0.95	115.8
Singapore	13.87	16.86	19.50	21.68	23.55	25.34	27.02	29.20	30.96	31.94	33.30	240.1
Thailand	0.59	0.65	0.72	0.79	0.80	0.89	0.94	1.03	1.21	1.66	1.70	288.1
Hong Kong	22.10	22.90	23.34	24.86	26.32	27.77	28.90	30.21	31.66	33.01	34.70	157.1
Japan	31.26	32.19	33.13	34.01	35.20	36.02	36.67	37.81	37.07	38.17	40.80	130.5
S. Korea	4.22	4.95	6.27	7.34	8.88	10.61	12.26	14.09	16.11	18.42	20.70	490.5
Sri Lanka	0.33	0.35	0.37	0.42	0.44	0.46	0.48	0.52	0.55	0.57	0.68	206.6
Iran	2.28	3.38	3.30	2.67	2.55	3.50	3.57	3.85	3.04	2.97	2.97	120.0
Israel	18.68	19.41	20.39	21.87	23.02	23.94	25.14	26.15	28.20	30.20	31.60	169.2
Pakistan	0.30	0.32	0.34	0.36	0.39	0.39	0.43	0.48	0.48	0.53	0.60	200.0
Syria	1.88	2.03	2.21	2.66	3.27	3.53	3.75	4.15	4.22	4.22	4.30	228.7
Turkey	2.00	2.28	2.46	2.57	2.87	3.24	3.48	4.02	4.37	5.35	7.00	350.0
UAE	11.95	10.81	10.81	11.60	12.71	13.95	14.51	15.26	13.27	14.29	18.20	152.3

Source: Data have been gathered from ITU, Yearbook of Common Carriers Telephone Statistics, 1973 and following years.

Chapter 6

CONCLUSIONS

Consensus is growing among economists that tremendous shifts have occurred in the speed and direction of technological change since the 1970s. A wave of innovations has flooded the economic system, affecting firms' strategies and industry's structure. Most of this change has originated in electronic innovations and their application in a spectrum of traditional activities.

The main features of this technological change are:

1. The technological leadership of the United States is declining. Japan and Europe are increasingly able to compete with the United States in many key technologies. This has important consequences for the international technology market and the variety and quantity of its goods[1];
2. The life cycle of innovations is short; therefore innovators must reap profits faster, sometimes by licensing their technology rather than by exporting it or establishing affiliates abroad[2];
3. Innovations introduced in the 1970s and 1980s can be used in many sectors and technologies totally unrelated to those of the innovator, and this intersectorial application by third parties keeps technology from being "appropriated" and favours "free riders"[3];
4. Innovations introduced in the 1970s and 1980s, such as advanced telecommunications, biotechnologies and informatics, are strongly systemic: they interrelate intensively with other innovated technologies and very little with existing capital stock. This offers many opportunities for fast-growing newly industrialised countries with high rates of investment and low levels of sunk costs;
5. Many process innovations of recent years can revitalise traditional manufacturing sectors, with more benefits for users than producers of the innovation. This, too, presents opportunities to developing countries in the international technology environment[4];
6. Increased competition on international markets among technological leaders is pushing monopolistic profits from innovations lower and lower, thus generating abundant consumer surplus for users in downstream industries[5].

Because of these changes, the conditions and timing of innovation diffusion both within and among countries is becoming central for assessing the role that technology will play in economic development in the next few years.

Early adopters of innovations gain an advantage in the international marketplace, where they can control strategic shares through exports of goods, "disembodied" technology (patents or licenses) and the growth of multinational firms. The international spread of information, however, narrows the technology gap between leaders and followers, allowing followers to buy technologies at lower marginal costs and compete effectively with leaders.

Economic analysis today seems more and more aware of the importance of information in understanding diffusion, and thus growth and development. Information and knowledge are imperfect – difficult to "own" and transfer on markets – but they are necessary for the acquisition and use of an innovation. Knowledge can take many forms; it can be obtained through learning by doing, learning by using, learning by imitating, and even "learning to learn".

Our micro-macroeconomic model of diffusion presents some insights not only into the economics of innovation diffusion but also the interaction of technological change, investment and productivity growth. Productivity growth, in our view, is strongly affected by the rate of diffusion, and more specifically by cognitive externalities like collective learning that "infuse innovation into the economic system"[6]. Because of bounded rationality concerning risk and lack of information about options, and the importance of knowledge for new technologies, the transfer of information on markets is costly and uncertain. Learning by imitating is much more effective.

In this perspective some "infusion" of new microeconomics based on assumptions of bounded rationality and imperfect knowledge appears to be essential to grasp the growth processes as shown by the neo-Keynesian approach and the connections between technological change, investment and productivity. When new "machines" are discontinuous, or incompatible, with existing ones and a learning process is necessary for their use, then diffusion lags due to bounded rationality and delayed adoption will result.

The empirical analysis of the international diffusion of the new information technologies (and particularly advanced telecommunications, which is heavily capital-intensive) has shown how macroeconomic factors are at least as influential as microeconomic elements such as collective learning and adoption profitability in the purchase of innovated capital goods. Macroeconomic conditions like the incentive to invest are essential for the cumulative growth process begun by the creation of an innovation. The chances for developing countries to take advantage of the benefits of technological change seem in fact rooted in the levels of investment they can sustain[7]. The macroeconomic approach also challenges the product-cycle concepts that diffusion among imitating countries takes place once it is complete within the innovating one; that innovators have high income, imitators low; and that exports from innovating countries and the presence of multinational firms are prerequisites for diffusion.

A macroeconomic theory of diffusion recognises the possibility of catching up and leapfrogging. Imitating countries can reach fast rates of diffusion, possibly faster rates than innovating countries, if they have access to efficient communication flows, good levels of cognitive and network externalities (such as collective learning and skilled manpower), and, especially, strong growth rates with consequent high levels of investment.

The evidence on the diffusion of advanced telecommunications confirms that:

1. Diffusion rates were particularly high in the years 1977-1987 compared to other key technological innovations diffused in recent years;
2. Diffusion rates during 1977-1987 were remarkably high in Far East Asia, and to a lesser extent, Latin America;
3. Diffusion was faster in this period in most late industrialising countries than in many early industrialised European countries;

4. Diffusion levels at the end of 1987 were unevenly distributed, with exceedingly high levels of penetration of digital and electronic switching in late industrialising countries with small markets, low wages, recent telecommunications infrastructure, and little existing vertical integration with telecommunications hardware manufacturers;

5. International diffusion of advanced telecommunications has had little to do with innovation capacity, which remains concentrated in a few early industrialised countries with the remarkable exception of Japan and (in the 1980s) a few small European countries;

6. The shake-up of the international telecommunications equipment industry by the introduction and application of microelectronics has completely altered supply and demand;

7. A sharp increase in the slope of long-term average cost curves owing to high R&D costs has raised barriers to new competitors but has improved mobility for existing manufacturers;

8. Technological discontinuity between electromechanical and electronic switching has erased the advantages of vertical integration, especially in small countries, diminishing the amount of captive markets and enhancing competition;

9. Changes in both supply and demand have increased international competition by pushing regional oligopolies to merge in an unstable global market, resulting in lower prices, cut-throat competitive strategies and "dumping" policies;

10. Highly regulated and government-protected telecommunications carriers, the customers of equipment suppliers, benefited from good bargaining positions and access to technological information. Because of this, in an epidemic approach to diffusion the fast adoption by late industrialising countries is not surprising;

11. Fast-growing countries, many of them in Far East Asia, experiencing high demand and investments could benefit from the heightened competition and buy at cheaper prices, soliciting "dumping" by suppliers through "soft loans" from home countries of large multinational companies;

12. In the 1980s international prices for electronic switching were often much lower than the domestic market price of many European countries that tried to sustain national high-tech industries with captive markets, "shadow prices" and export subsidies;

13. Investment decisions determined by rapid economic growth, poor infrastructure, low prices of telecommunications equipment and the equipment's increasing performance have been the leading factor in the fast rates of diffusion in late industrialising countries;

14. Good adoption capacity based on strong cognitive and network externalities like learning opportunities and expertise and large-sized firms that provide telecommunications services have quickly directed investment towards innovated equipment and resulted in fast rates of modernisation;

15. The transition from electromechanical switching to electronic switching, as in most network technologies, is deeply affected by the size and durability of the installed base of the incumbent technology and by the sustainable rate of growth of the installed base of the innovated technology.

In sum, the evidence of the international diffusion of advanced telecommunications confirms that the "appropriability" of innovations has greatly declined in recent years. Technological innovations introduced in upstream sectors of industry eventually induce fast diffusion in downstream sectors, and technology "blending" has spread throughout traditional activities.

The differing abilities of countries to adopt new technologies may become an important factor in the uneven distribution in international markets of expected productivity gains and economic growth. So far, international differences in diffusion rates have had a tremendous effect, directly and indirectly, on countries' competitiveness[8]. More explicitly it can be argued that a country's capacity to adopt new production processes, capital equipment and intermediary products rapidly is a good determinant of its performance on international markets, together with its traditional national and specific industry variables such as share of employment in multinational companies and existence of large firms. Diffusion rates are expected to be as important in assessing international competitiveness as the relative factor price variable measured by income per capita and capital intensity, or expenditure on R&D[9].

A strategy of fast adoption and use of process innovations both embodied and disembodied thus looks more promising today than a few years ago. According to the evidence, however, certain conditions are necessary. Fast diffusion rates of process innovations are in fact possible only with:

a) High levels of investment and growth;
b) Strong cognitive and network externalities generated by lead-users and critical mass;
c) Fast decline in the purchase prices of innovated capital goods;
d) Proximity to lead-producers such as multinational companies; and
e) Systemic technologies with low levels of sunk costs.

The rapid diffusion and adoption by imitating countries of advanced telecommunications highlights several key policy issues:

1. Bargaining power of procurement on the international markets of capital goods embodying technological innovations can be strengthened because of the globalisation of supply. National policies based on "staggered entry behaviour" (as in our example of Japan) on domestic markets and selected procurement capacity (as with Telecoms in Indonesia) can reinforce national imitators[10];
2. Fast adoption of key technologies in upstream sectors will probably create spillover effects, technological externalities and learning opportunities in downstream industries and increase overall productivity for the entire economic system;
3. A selective strategy of adopting key technologies and fast diffusion of advanced innovations and services is crucial to successful export-led growth strategies for the production of final consumer goods[11].

One final observation. The continuing debt burden on many developing countries imposes severe constraints on and may even exclude investments in basic infrastructure (such as advanced telecommunications) that is vital to enhanced productivity and competitiveness in downstream sectors. The result may be seriously to undermine not only a developing country's ability to continue servicing its debt, but also its ability to climb out of the debt trap by taking advantage of the new opportunities to compete internationally that are created by the technological revolution in telecommunications.

NOTES AND REFERENCES

1. M. Abramovitz, "Catching Up, Forging Ahead and Falling Behind", *Journal of Economic History* (June 1986), pp. 385-406.
2. L. Soete, "International Diffusion of Technology: Industrial Development and Technological Leapfrogging", *World Development* (March 1985), pp. 409-422.
3. K. Pavitt, "Sectoral Patterns of Technical Change: Towards a Taxonomy and a Theory", *Research Policy 6* (1984), pp. 343-373.
4. F.M. Scherer, "Interindustry Technology Flows in the U.S.", *Research Policy 5*, (1982), pp. 227-245.
5. F. Chesnais, "Science, Technology and Competitiveness", *Science, Technology and Industry Review 1* (1986), pp. 85-129.
6. "Whether the increase in output would be more or less proportionate to the increase in capital will depend [...] on the speed with which capital is accumulated, relatively to the capacity to innovate and to infuse innovation into the economic system. The more 'dynamic' are the people in control of production, the keener they are in search of improvements, and the readier they are to adopt new ideas and to introduce new ways of doing things, the faster production (per man) will rise, and the higher is the rate of accumulation of capital that can be maintained", N. Kaldor, "Capital Accumulation and Economic Growth" in V. Lutz, ed., *The Theory of Capital*, Macmillan, London, 1961, p. 36.
7. Our micro-macro model thus seems to give a new scope of application to the Kaldorian "technical progress function". See N. Kaldor and J.A. Mirrlees' "A New Model of Economic Growth", *Review of Economic Studies* (June 1962), pp. 174-192: "We shall assume that 'machines' of each vintage are of constant physical efficiency during their lifetime, so that the growth of productivity into the economy is entirely due to the infusion of new 'machines' into the system through (gross) investment. Hence our basic assumption is a technical progress function which makes the annual rate of growth of productivity per worker operating on a new equipment a function of the rate of growth of investment per worker".
8. C. Antonelli, "The International Diffusion of Process Innovations and the Neotechnology Theory of International Trade", *Economic Notes 1* (1986), pp. 140-161.
9. For a clear analysis of the relevance of slow diffusion rates in the economic decline of an industrialised country like the United Kingdom see K. Pavitt, ed., *Technological Innovation and British Economic Performance*, Macmillan, London, 1980.
10. T. Ozawa, "Government Control over Technology Acquisition and Firms' Entry into New Sectors: The Experience of Japan's Synthetic Fibre Industry", *Cambridge Journal of Economics* (June 1980), pp. 133-146.
11. High levels of investments combined with high levels of adoption capacity are instrumental in the creation of "virtuous cycles"; especially in the fast adoption of foreign technology. A fiscal policy aimed at increasing investment capacity is a basic tool of technology policy that an imitating country cannot afford to miss.

ANNEX I

A variety of logistic models have been elaborated in diffusion analysis literature. Davies (1979) has made an interesting distinction between the cumulative normal pattern, approximated by the logistic curve, and the cumulative log normal curve. The latter, with an inflection point at 37 per cent of penetration levels, would more appropriately fit the diffusion pattern of product innovations, while the former (with an inflection point at 50 per cent) would fit the diffusion pattern of more complex process innovations with longer learning times.

The logistic curve can be specified as follows:

[ii] $\qquad \log y_{it} = a + b_i \log t$

where $\qquad y_{it} = \dfrac{P_{it}}{K - P_{it}}$

P_{it} = the percentage of adopters compared to total potential adopters in country i at year t
K = the ceiling level of potential adopters set equal to 100.

Equation ii can also test for data of Table 4 in the period 1977-1986. Results of GLS estimates are in Table 17.

Equation ii performs better in a smaller number of countries (16) out of the 36 considered. The close performance of equation i (as seen in Chapter 3) and equation ii, with such large differences in diffusion patterns, suggests that the diffusion of advanced telecommunications has been so fast for so many countries in the early stages that it can be considered close to, and in many cases absolutely, a product innovation. The limited number of decision makers (usually monopolistic telecommunications carriers) involved in the adoption of central exchanges based on electronic switching helps to explain why, once the diffusion process has begun, it proceeds more quickly than that of complex process innovations that follow a more symmetric S-shaped pattern.

Table 17. Results of the GLS estimates of equation *ii*

	a	log b	$\bar{R}^2$
Canada	−6.411 (12.565)	2.295 (11.081)	0.924
USA	−8.72 (26.622)	3.479 (26.164)	0.985
Costa Rica	−27.916 (4.138)	8.459 (3.089)	0.460
El Salvador	−35.177 (7.433)	12.262 (6.383)	0.799
Mexico	−8.285 (5.619)	2.016 (3.369)	0.508
Panama	−25.478 (5.927)	9.106 (5.219)	0.724
Argentina	−25.597 (7.392)	8.413 (5.985)	0.776
Chile	−37.182 0(7.751)	13.147 (6.600)	0.809
Colombia	−26.864 (7.349)	9.628 (6.489)	0.804
Ecuador	−32.688 (10.712)	11.269 (9.097)	0.891
Peru	−32.170 (6.904)	11.763 (6.219)	0.790
Venezuela	−28.618 (9.904)	9.371 (7.990)	0.862
Morocco	−37.506 (5.823)	13.136 (5.024)	0.708
S. Africa	−11.730 (18.376)	3.411 (13.158)	0.945
Belgium	−18.202 (3.869)	6.739 (3.529)	0.533
Denmark	−9.451 (5.948)	2.515 (3.899)	0.586
Finland	−32.362 (8.065)	10.791 (6.625)	0.810
France	−14.822 (50.241)	5.553 (46.373)	0.995
Germany	−19.589 (7.579)	5.762 (5.492)	0.746
Ireland	−36.273 (6.119)	12.533 (5.209)	0.732
Italy	−30.082 (6.709)	9.650 (5.302)	0.730
Spain	−25.492 (9.454)	8.710 (7.958)	0.861
Sweden	−31.614 (3.816)	9.401 (3.970)	0.702

Table 17 *(cont.)*

	a	log b	$\bar{R}^2$
Switzerland	−32.015 (5.081)	10.414 (4.072)	0.609
United Kingdom	−7.263 (8.279)	2.519 (7.075)	0.830
Malta	−4.282 (11.943)	1.622 (11.148)	0.925
Malaysia	−8.135 (14.497)	2.996 (13.154)	0.945
New Zealand	−33.262 (8.249)	11.271 (6.886)	0.822
Philippines	−21.284 (19.484)	7.626 (17.198)	0.967
Singapore	−11.448 (12.167)	4.343 (11.332)	0.927
Thailand	−36.304 (6.249)	12.516 (5.308)	0.731
Hong Kong	−6.627 (17.579)	2.445 (15.981)	0.962
Japan	−34.108 (4.801)	11.397 (3.952)	0.593
S. Korea	−36.459 (10.596)	13.535 (9.691)	0.902
Sri Lanka	−35.189 (8.961)	11.618 (6.411)	0.898
Israel	−34.439 (9.606)	12.069 (8.294)	0.871
Pakistan	−20.847 (5.531)	6.556 (4.285)	0.634
Turkey	−33.936 (8.066)	11.555 (6.766)	0.817

ANNEX II

It seems useful to look at the results of the approach we used in Chapter 5 compared to those drawn from more traditional methods.

Certain refinements of the epidemic model, in fact, give an alternative specification to test our hypothesis. Following Chow (1967) and Stoneman (1976) we can apply the stock adjustment approach with the Gomperz curve. This approach tests the effects of investment on the growth rate of the stock of electronic lines, under the control of the lagged stock.

The epidemic diffusion process can be formulated as follows, between the existing stock Y and the desired or equilibrium level Y^+:

[15] $\quad \dfrac{dY}{dt} = a\, Y\, (\log Y^+ - \log Y)$

The differential equation yields, after logarithmic transformation, the following specification of the Gomperz curve:

[16] $\quad \log y_t - \log y_{t-1} = a\, (\log y^+ - \log y_{t-1})$

If the equilibrium stock can be written itself as a function of other variables in our case of INV (Investments) with constant elasticities:

[17] $\quad \log y^+ = b \log INV$

we can rewrite equation 17 with $\log y^+$ replaced by the function carrying the variable INV which affects its level. Our equation 17 thus becomes:

[18] $\quad \log y_t - \log y_{t-1} = a\, b + a\, b_1 \log INV - a \log y_{t-1}$

Econometric specification of equation 18 leads to equation 19:

[19] $\quad \log LE(t) - \log LE(t-1) = a_1 + a_2 b_1 \log INV(t) - a_3 \log LE_{(t-1)}$

where the variables are defined as:

LE(t) = total number of electronic switching lines installed in each country in each year;

INV = investment as measured by total investments in telecommunications (excluding building and land) in US$ per capita in each country in each year.

All log variables are in natural logarithms and the symbols are the expected ones.

The test of the model has been made with Generalised Least Squares (GLS) because of the serial correlation of the donor term detected by low levels of D.W. The results of these best with the most performing lags of INV of the test of equation 19 are given in Table 18 for each country.

The results strongly confirm our hypothesis and show that investment is very important to the diffusion of electronic switching across most countries. In terms of the values of the estimates, investment has been especially significant in newly industrialising countries of Latin America, such as Costa Rica, Mexico and Colombia; most late adopters of East Asia, like Korea, Thailand, Sri Lanka and Singapore; and the European countries Ireland, Italy, Switzerland and Finland.

The investment variable is quite weak in early adopters like the United Kingdom, France, Malaysia and Hong Kong.

The investment variable enters the test of equation 19 with the wrong sign only in the United States.

In sum, the stock adjustment approach elaborated by Chow (1967) and Stoneman (1967) gives strong evidence of the power of investments in the diffusion of electronic switching throughout countries, and confirms the results of our approach.

Table 18. Results of GLS estimates of equation 19

Country		constant		log LE(t-1)		log INV(t-k)		R^2
Canada	log LE(t) − log LE(t-1) =	3.160 (1.329)	−	0.099 (1.054) log LE(t-1)	+	0.419 (0.874)	log INV(t-3)	R^2 = 0.185
USA	log LE(t) − log LE(t-1) =	4.034 (2.345)	−	0.197 (3.397) log LE(t-1)	−	0.202 (1.074)	log INV(t-3)	R^2 = 0.888
Costa Rica	log LE(t) − log LE(t-1) =	27.188 (1.881)	+	0.495 (1.274) log LE(t-1)	+	6.573 (1.793)	log INV(t-3)	R^2 = 0.356
Mexico	log LE(t) − log LE(t-1) =	20.419 (3.708)	−	0.804 (3.268) log LE(t-1)	+	1.487 (4.152)	log INV(t-2)	R^2 = 0.714
Colombia	log LE(t) − log LE(t-1) =	16.864 (1.864)	−	0.579 (2.307) log LE(t-1)	+	1.166 (1.272)	log INV(t-3)	R^2 = 0.471
Finland	log LE(t) − log LE(t-1) =	38.751 (3.000)	−	0.069 (0.863) log LE(t-1)	+	12.213 (2.951)	log INV(t-3)	R^2 = 0.554
France	log LE(t) − log LE(t-1) =	4.043 (5.431)	−	0.150 (5.087) log LE(t-1)	+	0.246 (1.332)	log INV(t-2)	R^2 = 0.802
Germany	log LE(t) − log LE(t-1) =	24.336 (2.517)	−	0.882 (2.855) log LE(t-1)	+	3.529 (1.978)	log INV(t-2)	R^2 = 0.616
Ireland	log LE(t) − log LE(t-1) =	18.241 (2.513)	−	0.415 (2.071) log LE(t-1)	+	3.951 (2.345)	log INV(t-3)	R^2 = 0.450
Italy	log LE(t) − log LE(t-1) =	20.964 (1.601)	−	0.181 (1.020) log LE(t-1)	+	5.511 (1.537)	log INV(t-3)	R^2 = 0.254
Spain	log LE(t) − log LE(t-1) =	12.435 (1.846)	−	0.256 (2.196) log LE(t-1)	+	2.177 (1.191)	log INV(t-2)	R^2 = 0.457
Sweden	log LE(t) − log LE(t-1) =	11.416 (2.494)	−	0.390 (2.257) log LE(t-1)	+	1.629 (2.186)	log INV(t)	R^2 = 0.485
Switzerland	log LE(t) − log LE(t-1) =	19.972 (1.686)	−	0.103 (0.557) log LE(t-1)	+	7.364 (1.610)	log INV(t-3)	R^2 = 0.273
United Kingdom	log LE(t) − log LE(t-1) =	9.938 (2.268)	−	0.397 (2.312) log LE(t-1)	+	0.536 (1.129)	log INV(t-3)	R^2 = 0.433
Philippines	log LE(t) − log LE(t-1) =	4.193 (3.164)	−	0.229 (3.403) log LE(t-1)	+	0.054 (0.327)	log INV(t-2)	R^2 = 0.625
Singapore	log LE(t) − log LE(t-1) =	11.076 (3.648)	−	0.472 (4.024) log LE(t-1)	+	0.846 (2.525)	log INV(t-3)	R^2 = 0.724
Thailand	log LE(t) − log LE(t-1) =	56.687 (2.375)	−	0.310 (1.809) log LE(t-1)	+	7.487 (2.331)	log INV(t-3)	R^2 = 0.438
Hong Kong	log LE(t) − log LE(t-1) =	4.174	−	0.173 log LE(t-1)	+	0.223	log INV(t)	R^2 = 0.852

(1.475)   (1.029)   (1.331)

Malaysia $\log LE(t) - \log LE(t-1) = 9.108 - 0.401 \log LE(t-1) + 0.526 \log INV(t-2)$ $R^2 = 0.860$
     (4.256)   (4.558)   (2.971)

S. Korea $\log LE(t) - \log LE(t-1) = 38.080 - 0.791 \log LE(t-1) + 6.593 \log INV(t)$ $R^2 = 0.536$
     (2.473)   (2.577)   (2.289)

Sri Lanka $\log LE(t) - \log LE(t-1) = 36.640 - 0.884 \log LE(t-1) + 3.454 \log INV(t)$ $R^2 = 0.521$
     (1.455)   (2.037)   (1.215)

BIBLIOGRAPHY

ABRAMOVITZ, M., "Catching Up, Forging Ahead and Falling Behind", *Journal of Economic History* (June 1986), pp. 385-406.

ABRAMOVITZ, M., "Following and Leading," in H. Hanusch, ed., *"Evolutionary Economics Applications of Schumpeter's Ideas"*, Cambridge University Press, Cambridge, 1988.

ABRAMOVITZ, M., *Thinking about Growth*, Cambridge University Press, Cambridge, 1989.

ALLEN, D., "New Telecommunications Services: Network Externalities and Critical Mass", *Telecommunications Policy* (September 1988), pp. 257-271.

AMENDOLA, M. and J.L. GAFFARD, *The Innovative Choice: An Economic Analysis of the Dynamics of Technology*, Basil Blackwell, Oxford, 1988.

AMES, E., "Research Invention, Development and Innovation", *American Economic Review* (June 1961), pp. 370-381.

AMES, E., and N. ROSENBERG, "Changing Technological Leadership and Industrial Growth", *Economic Journal* (March 1963), pp. 13-31.

ANTONELLI, C., "Multinational Firms, International Trade and International Telecommunications", *Information Economics and Policy 4* (1984), pp. 333-343.

ANTONELLI, C., "The Diffusion of an Organizational Innovation: International Data Communications and Multinational Industrial Firms", *International Journal of Industrial Organization 2* (1985), pp. 109-118.

ANTONELLI, C., "The International Diffusion of New Information Technologies", *Research Policy 3* (1986), pp. 139-147.

ANTONELLI, C., "The International Diffusion of Process Innovations and the Neotechnology Theory of International Trade", *Economic Notes 1* (1986), pp. 60-82.

ANTONELLI, C., ed., *New Information Technology and Industrial Change: The Italian Case*, Kluwer Academic, Dordrecht and Boston, 1988.

ANTONELLI, C., "The Diffusion of Information Technology and the Demand for Telecommunications Services", *Telecommunications Policy* (September 1989), pp. 255-264.

ANTONELLI, C., "The Role of Technological Expectations in a Mixed Model of International Diffusion of Process Innovations: The Case of Open-End Spinning Rotors", *Research Policy 5* (1989), pp. 273-288.

ANTONELLI, C., "Information Technology and the Derived Demand for Telecommunications Services in the Manufacturing Industry", *Information Economics and Policy*, (1989/90), p.p. 45-55.

ANTONELLI, C., "Profitability and Imitation in the Diffusion of Process Innovations", *Rivista Internazionale di Scienze Economiche e Commerciali* (February 1990), pp. 109-126.

ANTONELLI, C., "Induced Adoption and Externalities in the Regional Diffusion of New Information Technology", *Regional Studies* (February 1990), pp. 31-40.

ANTONELLI, C., and L. GHEZZI, "Un'analisi teorica dei processi di diffusione dell'innovazione tecnologica in regime di monopolio temporaneo", *Giornale degli Economisti e Annali di Economia 3* (1987), pp. 125-148.

ANTONELLI, C. and G. GOTTARDI, "Interazioni tra produttori e utilizzatori nei processi di diffusione tecnologica", *L'Industria 4* (1988), pp. 599-636.

ANTONELLI, C., P. PETIT and G. TAHAR, "La Diffusion d'une nouvelle technique: L'application d'une innovation dans l'industrie textile", *Revue d'Economie Industrielle 48* (1989), pp. 1-15.

ANTONELLI, C., P. PETIT and G. TAHAR, "Technological Diffusion and Investment Behavior: The Case of the Textile Industry", *Weltwirtschaftliches Archiv 4* (1989), pp. 782-803.

ANTONELLI, C., P. PETIT and G. TAHAR, "The Diffusion of Interdependent Innovations in the Textile Industry", *Structural Change and Economic Dynamics 2*, (1991), pp. 1-19.

ARROW, K.J., "The Economic Implications of Learning by Doing", *Review of Economic Studies* (June 1962), pp. 155-173.

ARROW, J.K., *Economics of Information*, Harvard University Press, Cambridge, 1984.

ARROW, J.K. and A.C. FISHER, "Environmental Preservation Uncertainty and Irreversibility", *Quarterly Journal of Economics* (May 1974), pp. 312-320.

AT&T, *The World Telephones*, Morristown, 1975 and following years.

BALCER, Y. and S.A. LIPPMAN, "Technological Expectations and Adoption of Improved Technology", *Journal of Economic Theory* (December 1984), pp. 292-318.

BENIGER, J.R., *The Control Revolution: Technological and Economic Origins of the Information Society*, Harvard University Press, Cambridge, 1986.

BHALLA, A.S. and D. JAMES, eds., *New Technologies and Development: Experiences in Technology Blending*, Lyenne Rienner Publishers, Boulder and London, 1988.

BHATTACHARYA, S., K. CHATTERJEE and L. SAMUELSON, "Sequential Research and the Adoption of Innovations", Oxford Economic Papers (November 1986), pp. 219-243.

BRADLEY, S.P. and J.A. HAUSMAN, eds., *Future Competition in Telecommunications*, Harvard Business School Press, Boston, 1989.

BROCK, G.W., *The Telecommunications Industry: The Dynamics of Market Structure*, Harvard University Press, Cambridge, 1981.

CAVES, R.E., "Vent-for-Surplus Models of Trade and Growth", in Essays in Honor of G. Heberler, *Trade Growth and the Balance of Payments*, Rand McNally, Chicago, 1965.

CAVES, R.E., *Multinational Enterprise and Economic Analysis*, Cambridge University Press, Cambridge, 1982.

CAVES, R.E. and M.E. PORTER, "From Entry Barriers to Mobility Barriers: Conjectural Decisions and Contrived Deterrence to New Competition", *Quarterly Journal of Economics* (May 1977), pp. 241-261.

CHARLES, D., P. MONK and E. SCIBERRAS, *Technology and Competition in the International Telecommunications Industry*, Frances Pinter, London, 1989.

CHEN, E.K.Y., "Multinational Corporations and Technology Diffusion in Hong Kong Manufacturing", *Applied Economics* (June 1983), pp. 309-321.

CHESNAIS, F., "Science, Technology and Competitiveness", *Science, Technology and Industry Review 1* (1986), pp. 85-129.

CHESNAIS, F., "Multination Enterprises and the International Diffusion of Technology", in DOSI *et al.*, eds., *Technological Change and Economic Theory*, Pinter Publishers, London, 1988.

CHOW, G.C., "Technological Change and the Demand for Computers", *American Economic Review* (December 1967), pp. 1117-1130.

CHUDNOVSKI, D., "La Industria de Equipos de Telecommunicationes en la Argentina", CET-IPAL, Buenos Aires, 1985.

CORNWALL, J., *Growth and Stability in a Mature Economy*, Martin Robertson, London, 1972.

CORNWALL, J., *Modern Capitalism: Its Growth and Transformation*, Martin Robertson, London, 1977.

COURVILLE, L., ed., *Economic Analysis of Telecommunications*, North Holland, Amsterdam, 1983.

DALIA, M., "Import Led Innovation: The Case of the Austrian Textile Industry", *Weltwirtschaftliches Archiv 3* (1988), pp. 550-565.

DANG NGUYEN, G., "Telecommunications: A Challenge to the Old Order", in M. Sharp, ed., *Europe and the New Technologies*, Frances Pinter, London, 1985.

DASGUPTA, P. and J.E. STIGLITZ, "Learning by Doing: Market Structure and Industrial and Trade Policies", *Oxford Economic Papers* (June 1988), pp. 246-268.

DAVID, P., "The Mechanization of Reaping in the Anti-Bellum Midwest", in N. Rosenberg, ed., *The Economics of Technological Change*, Penguin, Harmondsworth, 1971.

DAVID, P.A., *Technical Choice, Innovation and Economic Growth*, Cambridge University Press, Cambridge, 1975.

DAVID, P., "CLIO and the Economics of QWERTY", *American Economic Review* (May 1985), pp. 332-337.

DAVID, P.A., "Some New Standards for the Economics of Standardization in the Information Age", in P. Dasgupta and P. Stoneman, eds., *Economic Policy and Technological Performance*, Cambridge University Press, Cambridge, 1987.

DAVID, P.A. and J.A. BUNN, "The Evolution of Gateway Technologies and Network Evolution: Lessons from Electricity Supply History", *Information Economics and Policy 3* (1988), pp. 165-202.

DAVID, P.A. and T.E. OLSEN, "Equilibrium Dynamics of Diffusion when Incremental Innovations are Foreseen", *Stanford University Center for Economic Policy Research Working Paper no. 9*, 1986.

DAVIES, S., *The Diffusion of Process Innovations*, Cambridge University Press, London, 1979.

DAY, R.H., "Endogenous Preferences and Adoptive Economizing", in R.H. Day and G. Eliasson, eds., *The Dynamics of Market Economics*, North Holland, Amsterdam, 1986.

DEPT. OF TOURISM, POST AND TELECOMMUNICATIONS, *Strategic Development Plan*, Jakarta, 1989.

FABERGER, J., "A Technology-Gap Approach to Why Growth Rates Differ", *Research Policy* (August 1987), pp. 87-99.

FABERGER, J., "Why Growth Rates Differ", in DOSI *et al.*, eds., *Technical Change and Economic Theory*, Pinter Publishers, London, 1988.

FARRELL, J. and G. SALONER, "Standardization Compatibility and Innovation", *Rand Journal of Economics* (Spring 1985), pp. 70-83.

FARRELL, J. and G. SALONER, "Competition, Compatibility and Standards: The Economics of Horses, Penguins and Lemmings", in H.L. Gabel, ed., *Product Standardization as a Tool of Competitive Strategy*, North Holland, Amsterdam, 1988.

FRANKEL, M., "Obsolescence and Technical Change in a Maturing Economy", *American Economic Review* (September 1955), pp. 296-319.

FRANSMAN, M., *Beyond the Market: Cooperation and Competition in Information Technology in the Japanese System*, Cambridge University Press, Cambridge, 1990.

FREEMAN, C., *The Economics of Industrial Innovation*, 2nd edition, Pinter Publishers, London, 1982.

FREEMAN, C. and C. PEREZ, "Structural Crisis of Adjustment: Business Cycles and Investment Behaviour", in G. DOSI, *et al*, eds., *Technological Change and Economic Theory*, Pinter Publishers, London, 1988.

GRILICHES, Z., "Hybrid Corn: An Exploration in the Economics of Technological Change", *Econometrica* (October 1957), pp. 501-522.

GRILICHES, Z., ed., *R&D, Patents and Productivity*, University of Chicago Press for National Bureau of Economic Research, Chicago, 1984.

HAYEK, F.A., *Individualism and Economic Order*, University of Chicago Press, Chicago, 1948.

HERRERA, A., *La revolucion tecnologica y las telefonie argentina. De la Union Telefonica à la Telefonica Argentina*, Editorial Legaza, Buenos Aires, 1989.

HIRSCH, S., *Location of Industry and International Competitiveness*, Oxford University Press, London, 1967.

HIRSCHMAN, A.O., *The Strategy of Economic Development*, Yale University Press, New Haven, 1958.

HIRSCHMAN, A.O., "The Political Economy of Import-Substituting Industrialization in Latin America", *Quarterly Journal of Economics* (February 1968), pp. 1-33.

VON HIPPEL, E., *The Sources of Innovation*, Oxford University Press, Oxford, 1988.

HOBDAY, M.G., *Telecommunications and the Developing Countries: the Challenge from Brazil*, Routledge, London, 1989.

HSIA, R., "Technological Change in the Industrial Growth of Hong Kong", in B.R. Williams, ed., *Science and Technology in Economic Growth*, Macmillan, London, 1973.

HUDSON, H.H. and L.C. YORK, "Generating Foreign Exchange in Developing Countries: The Potential of Telecommunications Investments", *Telecommunications Policy* (September 1988), pp. 272-281.

ITALTEL, *L'annuario delle telecomunicazioni*, Milano, 1984 and following years.

ITU (International Telecommunications Union), *Yearbook of Common Carriers Telecommunications Statistics*, Geneva, 1973 and following years.

JECQUIER, N., "International Technology Transfer in the Telecommunications Industry", in D. Germidis, ed., *Transfer of Technology by Multinational Corporations*, OECD, Paris, 1977.

KALDOR, N., "A Model of Economic Growth", *Economic Journal*, (December 1957), pp. 591-624.

KALDOR, N., "Capital Accumulation and Economic Growth" in V. Lutz, ed., *The Theory of Capital*, Macmillan, London, 1961.

KALDOR, N., "Increasing Returns and Technical Progress: A Comment on Professor Hick's Article", *Oxford Economic Papers* (February 1961), pp. 1-4.

KALDOR, N., "Comment", *Review of Economics and Statistics 48*, (July 1962), pp. 118-120.

KALDOR, N., *Strategic Factors in Economic Development*, Cornell University, Ithaca, 1967.

KALDOR, N., "Productivity and Growth: A Reply", *Economica* (November 1968), pp. 385-391.

KALDOR, N., "The Irrelevance of Equilibrium Economics", *Economic Journal* (December 1972), pp. 1237-1255.

KALDOR, N. and J.A. MIRRLEES, "A New Model of Economic Growth", *Review of Economic Studies* (June 1962), pp. 174-192.

KARLSON, S.H., "Adoption of Competing Inventions by United States Steel Producers", *Review of Economics and Statistics* (August 1986), pp. 415-422.

KATZ, M.L. and C. SHAPIRO, "Network Externalities, Competition and Compatibility", *American Economic Review* (June 1985), pp. 424-440.

KATZ, R., *The Information Society: An International Perspective*, Praeger, New York, 1988.

LAMBORGHINI, B. and C. ANTONELLI, "The Impact of Electronics on Industrial Structures and Firms' Strategies", in OECD, *Microelectronics, Productivity and Employment,* Paris, 1981.

LEFF, N. H., "Social Benefit Cost-Analysis and Telecommunications Investment in Developing Countries", *Information Economics and Policy* 3 (1984), pp. 217-227.

LEKVALL, P. and C. WAHLBIN, "A Study on Some Assumptions Underlying Innovation Diffusion Functions", *Swedish Journal of Economics* (December 1973), pp. 362-377.

LOASBY, B.J, *Choice Complexity and Ignorance,* Cambridge University Press, Cambridge, 1976.

LYNN, L., "New Data on the Diffusion of the Basic Oxygen Furnace in the U.S. and Japan", *Journal of Industrial Economics* (December 1981), pp. 123-135.

MADDALA, G.S. and P.T. KNIGHT, "International Diffusion of Technical Change: A Case Study in the Oxygen Steel-Making Process", *Economic Journal* (September 1967), pp. 531-558.

MAHAJAN, V. and Y. WIND, eds., *Innovation Diffusion Models of New Product Acceptance,* Ballinger, Cambridge, 1986.

MANSFIELD, E., "Technical Change and the Rate of Imitation", *Econometrica* (October 1961), pp. 741-766.

MANSFIELD, E., *Industrial Research and Technological Innovation,* W.W. Norton, New York, 1968.

METCALFE, J.S., "Impulse and Diffusion in the Study of Technical Change", *Futures* (October 1981), pp. 347-359.

NABSETH, L. and G.F. RAY, eds., *The Diffusion of New Industrial Processes,* Cambridge University Press, Cambridge, 1974.

NELSON, R.R., "A Diffusion Model of International Productivity Differences in Manufacturing Industry", *American Economic Review* (December 1968), pp. 1219-1248.

NELSON, R.R., M.J. PECK and E. KALACHEK, *Technology, Economic Growth and Public Policy,* The Brookings Institution, Washington, D.C., 1967.

NELSON, R.R. and S.G. WINTER, *An Evolutionary Theory of Economic Change,* Belknap Press of Harvard University Press, Cambridge, 1982.

OECD, *Telecommunications: Pressures and Policies for Change,* Paris, 1983.

OECD, *The Telecommunications Industry: The Challenges of Structural Change,* Paris, 1988.

OZAWA, T., "Government Control over Technology Acquisition and Firms' Entry into New Sectors: The Experience of Japan's Synthetic Fibre Industry", *Cambridge Journal of Economics* (June 1980), pp. 133-146.

PAVITT, K., ed., *Technological Innovation and British Economic Performance,* Macmillan, London, 1980.

PAVITT, K., "R&D Patenting and Innovative Activities", *Research Policy 6* (1982), pp. 33-51.

PAVITT, K., "Sectoral Patterns of Technical Change: Towards a Taxonomy and a Theory", *Research Policy 6* (1984), pp. 343-373.

PERES NUNEZ, W., *Foreign Direct Investment and Industrial Development in Mexico,* OECD Development Centre, Paris, 1990.

PETIT, P. and G. TAHAR, "Dynamics of Technological Change and Schemes of Diffusion", *Manchester School of Economics and Social Studies* (December 1989), pp. 370-386.

POSNER, M.V., "International Trade and Technical Change", *Oxford Economic Papers* (October 1961), pp. 330-337.

RAY, G.F., "The Diffusion of New Technology: A Study in Ten Processes in Nine Industries", *National Institute Economic Review* (May 1969), pp. 40-83.

RAY, G.F., *The Diffusion of Mature Technologies,* Cambridge University Press, Cambridge, 1984.

RAY, G.F., "Full Circle: The Diffusion of Technology", *Research Policy 1* (1989), pp. 1-18.

REINGANUM, J.F., "Market Structure and the Diffusion of New Technology", *Bell Journal of Economics* (Autumn 1981), pp. 618-624.

ROGERS, E.M., *Diffusion of Innovations,* 3rd ed., Free Press, New York, 1983.

ROMER, P.M., "Increasing Returns and Long-Run Growth", *Journal of Political Economy* (October 1986), pp. 1002-1037.

ROOBECK, A.J., "Telecommunications: An Industry in Transition", in H.W. de Jong, ed., *The Structure of the European Industry,* Kluwer Academic Publishers, Dordrecht and Boston, 1988.

ROSENBERG, N., *Perspectives on Technology,* Cambridge University Press, Cambridge, 1976.

ROSENBERG, N., *Inside the Black Box: Technology and Economics,* Cambridge University Press, Cambridge, 1982.

ROSENBERG, N., "New Technology and Old Debates", in A.S. Bhalla and D. James, eds., *New Technology and Development: Experiences in "Technology Blending"* , Lyenne Rienner Publishers, Boulder and London, 1988.

SALTER, W.E.G., *Productivity and Technical Change,* Cambridge University Press, Cambridge, 1960.

SCHERER, F.M., "Interindustry Technology Flows in the U.S.", *Research Policy* 5 (1982), pp. 227-245.

SCOTT, M.F.G., *A New View of Economic Growth,* Clarendon Press, Oxford, 1989.

SIEMENS, *International Telephone Statistics,* Munich, 1984 and following years.

SIMON, H.A., *Models of Bounded Rationality: Behavioral Economics and Business Organization,* MIT Press, Cambridge, 1982.

SOETE, L., "International Diffusion of Technology: Industrial Development and Technological Leapfrogging", *World Development* (March 1985), pp. 409-422.

SOUNDERS, R.J., J.J. WARFORD and B. WELLENIUS, *Telecommunications and Economic Development,* Johns Hopkins University Press, Baltimore, 1983.

STERN, N., "The Economics of Development: A Survey", *Economic Journal* (September 1989), pp. 597-685.

STIGLITZ, J.E., "Learning to Learn: Localized Learning and Technological Progress", in P. Dasgupta and P. Stoneman, eds., *Economic Policy and Technological Performance,* Cambridge University Press, London, 1987.

STIGLITZ, J.E., "Economic Organization, Information and Development", in H. Chenery and T.S. Srinivasan, eds., *Handbook of Development Economics Vol. 1,* Elsevier, Amsterdam, 1988.

STIGLITZ, J.E., "Markets, Market Failures and Development", *American Economic Review* (May 1989), pp. 197-203.

STOBAUGH, R., *Innovation and Competition: The Global Management of Petrochemical Products,* Harvard Business School University Press, Boston, 1988.

STONEMAN, P., *Technological Diffusion and the Computer Revolution,* Cambridge University Press, Cambridge, 1976.

STONEMAN, P., *The Economic Analysis of Technological Change,* Oxford University Press, London, 1983.

STONEMAN, P., *The Economic Analysis of Technology Policy,* Clarendon Press, Oxford, 1987.

STONEMAN, P. and N.J. IRELAND, "The Role of Supply Factors in the Diffusion of New Process Technology", *Economic Journal* (March 1983), pp. 66-78.

STREISSLER, E., "Growth Models As Diffusion Processes", *Kyklos* 1/2 (1979), pp. 251-269.

SUTHERLAND, A., "The Diffusion of an Innovation in Cotton Spinning", *Journal of Industrial Economics* (March 1959), pp. 118-135.

SUTTON, C.J., "The Effect of Uncertainty on the Diffusion of Third Generation Computers", *Journal of Industrial Economics* (June 1975), pp. 273-280.

SWANN, P.L., "The International Diffusion of Innovation", *Journal of Industrial Economics* (September 1973), pp. 61-69.

TELECOMMUNICATIONS AUTHORITY OF SINGAPORE, *Telecoms Annual Report*, Singapore 1982 and following years.

THIRTLE, C.G. and V.W. RUTTAN, *The Role of Demand and Supply in the Generation and Diffusion of Technical Change*, Harwood Academic Publishers, Chur, 1987.

TILTON, J., *International Diffusion of Technology, The Case of Semiconductors*, The Brookings Institution, Washington, 1971.

UNGERER, H., *Telecommunications in Europe*, C.E.C., Brussels, 1988.

VERNON, R., "International Investment and International Trade in the Product Cycle", *Quarterly Journal of Economics* (May 1966), pp. 198-207.

WILLIAMSON, O.E., *Markets and Hierarchies: Analysis and Anti-Trust Implications*, Free Press, New York, 1975.

WILLIAMSON, O.E., *The Economic Institutions of Capitalism*, Free Press, New York, 1985.

YOUNG, A., "Increasing Returns and Economic Progress", *Economic Journal* (December 1982), pp. 527-542.

WHERE TO OBTAIN OECD PUBLICATIONS – OÙ OBTENIR LES PUBLICATIONS DE L'OCDE

Argentina – Argentine
CARLOS HIRSCH S.R.L.
Galería Güemes, Florida 165, 4° Piso
1333 Buenos Aires   Tel. 30.7122, 331.1787 y 331.2391
Telegram: Hirsch-Baires
Telex: 21112 UAPE-AR. Ref. s/2901
Telefax:(1)331-1787

Australia – Australie
D.A. Book (Aust.) Pty. Ltd.
648 Whitehorse Road, P.O.B 163
Mitcham, Victoria 3132   Tel. (03)873.4411
Telefax: (03)873.5679

Austria – Autriche
OECD Publications and Information Centre
Schedestrasse 7
D-W 5300 Bonn 1 (Germany)   Tel. (49.228)21.60.45
Telefax: (49.228)26.11.04
Gerold & Co.
Graben 31
Wien I   Tel. (0222)533.50.14

Belgium – Belgique
Jean De Lannoy
Avenue du Roi 202
B-1060 Bruxelles   Tel. (02)538.51.69/538.08.41
Telex: 63220   Telefax: (02) 538.08.41

Canada
Renouf Publishing Company Ltd.
1294 Algoma Road
Ottawa, ON K1B 3W8   Tel. (613)741.4333
Telex: 053-4783   Telefax: (613)741.5439
Stores:
61 Sparks Street
Ottawa, ON K1P 5R1   Tel. (613)238.8985
211 Yonge Street
Toronto, ON M5B 1M4   Tel. (416)363.3171
Federal Publications
165 University Avenue
Toronto, ON M5H 3B8   Tel. (416)581.1552
Telefax: (416)581.1743
Les Publications Fédérales
1185 rue de l'Université
Montréal, PQ H3B 3A7   Tel.(514)954-1633
Les Éditions La Liberté Inc.
3020 Chemin Sainte-Foy
Sainte-Foy, PQ G1X 3V6   Tel. (418)658.3763
Telefax: (418)658.3763

Denmark – Danemark
Munksgaard Export and Subscription Service
35, Nørre Søgade, P.O. Box 2148
DK-1016 København K   Tel. (45 33)12.85.70
Telex: 19431 MUNKS DK   Telefax: (45 33)12.93.87

Finland – Finlande
Akateeminen Kirjakauppa
Keskuskatu 1, P.O. Box 128
00100 Helsinki   Tel. (358 0)12141
Telex: 125080   Telefax: (358 0)121.4441

France
OECD/OCDE
Mail Orders/Commandes par correspondance:
2, rue André-Pascal
75775 Paris Cédex 16   Tel. (33-1)45.24.82.00
Bookshop/Librairie:
33, rue Octave-Feuillet
75016 Paris   Tel. (33-1)45.24.81.67
(33-1)45.24.81.81
Telex: 620 160 OCDE
Telefax: (33-1)45.24.85.00 (33-1)45.24.81.76
Librairie de l'Université
12a, rue Nazareth
13100 Aix-en-Provence   Tel. 42.26.18.08
Telefax : 42.26.63.26

Germany – Allemagne
OECD Publications and Information Centre
Schedestrasse 7
D-W 5300 Bonn 1   Tel. (0228)21.60.45
Telefax: (0228)26.11.04

Greece – Grèce
Librairie Kauffmann
28 rue du Stade
105 64 Athens   Tel. 322.21.60
Telex: 218187 LIKA Gr

Hong Kong
Swindon Book Co. Ltd.
13 - 15 Lock Road
Kowloon, Hong Kong   Tel. 366.80.31
Telex: 50 441 SWIN HX   Telefax: 739.49.75

Iceland – Islande
Mál Mog Menning
Laugavegi 18, Pósthólf 392
121 Reykjavik   Tel. 15199/24240

India – Inde
Oxford Book and Stationery Co.
Scindia House
New Delhi 110001   Tel. 331.5896/5308
Telex: 31 61990 AM IN
Telefax: (11)332.5993
17 Park Street
Calcutta 700016   Tel. 240832

Indonesia – Indonésie
Pdii-Lipi
P.O. Box 269/JKSMG/88
Jakarta 12790   Tel. 583467
Telex: 62 875

Ireland – Irlande
TDC Publishers – Library Suppliers
12 North Frederick Street
Dublin 1   Tel. 744835/749677
Telex: 33530 TDCP EI   Telefax: 748416

Italy – Italie
Libreria Commissionaria Sansoni
Via Benedetto Fortini, 120/10
Casella Post. 552
50125 Firenze   Tel. (055)64.54.15
Telex: 570466   Telefax: (055)64.12.57
Via Bartolini 29
20155 Milano   Tel. 36.50.83
La diffusione delle pubblicazioni OCSE viene assicurata
dalle principali librerie ed anche da:
Editrice e Libreria Herder
Piazza Montecitorio 120
00186 Roma   Tel. 679.46.28
Telex: NATEL I 621427
Libreria Hoepli
Via Hoepli 5
20121 Milano   Tel. 86.54.46
Telex: 31.33.95   Telefax: (02)805.28.86
Libreria Scientifica
Dott. Lucio de Biasio 'Aeiou'
Via Meravigli 16
20123 Milano   Tel. 805.68.98
Telefax: 800175

Japan – Japon
OECD Publications and Information Centre
Landic Akasaka Building
2-3-4 Akasaka, Minato-ku
Tokyo 107   Tel. (81.3)3586.2016
Telefax: (81.3)3584.7929

Korea – Corée
Kyobo Book Centre Co. Ltd.
P.O. Box 1658, Kwang Hwa Moon
Seoul   Tel. (REP)730.78.91
Telefax: 735.0030

Malaysia/Singapore – Malaisie/Singapour
Co-operative Bookshop Ltd.
University of Malaya
P.O. Box 1127, Jalan Pantai Baru
59700 Kuala Lumpur
Malaysia   Tel. 756.5000/756.5425
Telex: 757.3661
Information Publications Pte. Ltd.
Pei-Fu Industrial Building
24 New Industrial Road No. 02-06
Singapore 1953   Tel. 283.1786/283.1798
Telefax: 284.8875

Netherlands – Pays-Bas
SDU Uitgeverij
Christoffel Plantijnstraat 2
Postbus 20014
2500 EA's-Gravenhage   Tel. (070 3)78.99.11
Voor bestellingen: Tel. (070 3)78.98.80
Telex: 32486 stdru   Telefax: (070 3)47.63.51

New Zealand – Nouvelle-Zélande
GP Publications Ltd.
Customer Services
33 The Esplanade - P.O. Box 38-900
Petone, Wellington
Tel. (04)685-555   Telefax: (04)685-333

Norway – Norvège
Narvesen Info Center - NIC
Bertrand Narvesens vei 2
P.O. Box 6125 Etterstad
0602 Oslo 6   Tel. (02)57.33.00
Telex: 79668 NIC N   Telefax: (02)68.19.01

Pakistan
Mirza Book Agency
65 Shahrah Quaid-E-Azam
Lahore 3   Tel. 66839
Telex: 44886 UBL PK. Attn: MIRZA BK

Portugal
Livraria Portugal
Rua do Carmo 70-74, Apart. 2681
1117 Lisboa Codex   Tel.: 347.49.82/3/4/5
Telefax: (01) 347.02.64

Singapore/Malaysia – Singapour/Malaisie
See Malaysia/Singapore" – Voir «Malaisie/Singapour»

Spain – Espagne
Mundi-Prensa Libros S.A.
Castelló 37, Apartado 1223
Madrid 28001   Tel. (91) 431.33.99
Telex: 49370 MPLI   Telefax: 575.39.98
Libreria Internacional AEDOS
Consejo de Ciento 391
08009 - Barcelona   Tel. (93) 301-86-15
Telefax: (93) 317-01-41
Llibreria de la Generalitat
Palau Moja, Rambla dels Estudis, 118
08002 - Barcelona   Telefax: (93) 412.18.54
Tel. (93) 318.80.12 (Subscripcions)
(93) 302.67.23 (Publicacions)

Sri Lanka
Centre for Policy Research
c/o Mercantile Credit Ltd.
55, Janadhipathi Mawatha
Colombo 1   Tel. 438471-9, 440346
Telex: 21138 VAVALEX CE   Telefax: 94.1.448900

Sweden – Suède
Fritzes Fackboksföretaget
Box 16356, Regeringsgatan 12
103 27 Stockholm   Tel. (08)23.89.00
Telex: 12387   Telefax: (08)20.50.21
Subscription Agency/Abonnements:
Wennergren-Williams AB
Nordenflychtsvägen 74, Box 30004
104 25 Stockholm   Tel. (08)13.67.00
Telex: 19937   Telefax: (08)618.62.32

Switzerland – Suisse
OECD Publications and Information Centre
Schedestrasse 7
D-W 5300 Bonn 1 (Germany)   Tel. (49.228)21.60.45
Telefax: (49.228)26.11.04
Librairie Payot
6 rue Grenus
1211 Genève 11   Tel. (022)731.89.50
Telex: 28356
Subscription Agency – Service des Abonnements
Naville S.A.
7, rue Lévrier
1201 Genève   Tél.: (022) 732.24.00
Telefax: (022) 738.48.03
Maditec S.A.
Chemin des Palettes 4
1020 Renens/Lausanne   Tel. (021)635.08.65
Telefax: (021)635.07.80
United Nations Bookshop/Librairie des Nations-Unies
Palais des Nations
1211 Genève 10   Tel. (022)734.14.73
Telex: 412962   Telefax: (022)740.09.31

Taiwan – Formose
Good Faith Worldwide Int'l. Co. Ltd.
9th Floor, No. 118, Sec. 2
Chung Hsiao E. Road
Taipei   Tel. 391.7396/391.7397
Telefax: (02) 394.9176

Thailand – Thaïlande
Suksit Siam Co. Ltd.
1715 Rama IV Road, Samyan
Bangkok 5   Tel. 251.1630

Turkey – Turquie
Kültur Yayinlari Is-Türk Ltd. Sti.
Atatürk Bulvari No. 191/Kat. 21
Kavaklidere/Ankara   Tel. 25.07.60
Dolmabahce Cad. No. 29
Besiktas/Istanbul   Tel. 160.71.88
Telex: 43482B

United Kingdom – Royaume-Uni
HMSO
Gen. enquiries   Tel. (071) 873 0011
Postal orders only:
P.O. Box 276, London SW8 5DT
Personal Callers HMSO Bookshop
49 High Holborn, London WC1V 6HB
Telex: 297138   Telefax: 071 873 2000
Branches at: Belfast, Birmingham, Bristol, Edinburgh,
Manchester

United States – États-Unis
OECD Publications and Information Centre
2001 L Street N.W., Suite 700
Washington, D.C. 20036-4910   Tel. (202)785.6323
Telefax: (202)785.0350

Venezuela
Libreria del Este
Avda F. Miranda 52, Aptdo. 60337, Edificio Galipán
Caracas 106   Tel. 951.1705/951.2307/951.1297
Telegram: Libreste Caracas

Yugoslavia – Yougoslavie
Jugoslovenska Knjiga
Knez Mihajlova 2, P.O. Box 36
Beograd   Tel.: (011)621.992
Telex: 12466 jk bgd   Telefax: (011)625.970

Orders and inquiries from countries where Distributors
have not yet been appointed should be sent to: OECD
Publications Service, 2 rue André-Pascal, 75775 Paris
Cedex 16, France.
Les commandes provenant de pays où l'OCDE n'a pas
encore désigné de distributeur devraient être adressées à :
OCDE, Service des Publications, 2, rue André-Pascal,
75775 Paris Cédex 16, France.

75880-7/91

OECD PUBLICATIONS, 2 rue André-Pascal, 75775 PARIS CEDEX 16
PRINTED IN FRANCE
(41 91 13 1) ISBN 92-64-13578-2 - No. 45679 1991